Applied Principles of
Electrical Engineering

Applied Principles of Electrical Engineering

Edited by
Hope Miller

WILLFORD PRESS

www.willfordpress.com

Published by Willford Press,
118-35 Queens Blvd., Suite 400,
Forest Hills, NY 11375, USA

ISBN: 978-1-68285-757-1

Cataloging-in-Publication Data

Applied principles of electrical engineering / edited by Hope Miller.
  p. cm.
Includes bibliographical references and index.
ISBN 978-1-68285-757-1
1. Electrical engineering. 2. Electrical engineering--Materials. 3. Engineering. I. Miller, Hope.
TK145 .A66 2020
621.3--dc23

For information on all Willford Press publications
visit our website at www.willfordpress.com

WILLFORD PRESS

Contents

Preface

An engineering discipline, which is concerned with the study and application of electricity, electronics and electromagnetism, is called electrical engineering. Various fields that fall under the domain of electrical engineering include computer engineering, electronics, control systems, signal processing, digital computers, etc. An important sub-field of electrical engineering is power engineering. It is concerned with the generation, distribution and transmission of electricity, along with the design of a range of related devices. These include electric motors, electric generators, transformers and power electronics. The field of telecommunication focuses on the transmission of information across a communication channel such as free space, coax cable and optical fiber. The information has to be encoded in a carrier signal to transfer it to a carrier frequency that is suitable for transmission. The use of nanotechnology in electrical engineering has resulted in the production of nanoelectronics. It covers a diverse set of devices and materials such as one-dimensional nanotubes or nanowires and advanced molecular electronics. The various sub-fields of electrical engineering along with the technological progress that has future implications are glanced at in this book. It will serve as a valuable source of reference for graduate and postgraduate students.

Various studies have approached the subject by analyzing it with a single perspective, but the present book provides diverse methodologies and techniques to address this field. This book contains theories and applications needed for understanding the subject from different perspectives. The aim is to keep the readers informed about the progresses in the field; therefore, the contributions were carefully examined to compile novel researches by specialists from across the globe.

Indeed, the job of the editor is the most crucial and challenging in compiling all chapters into a single book. In the end, I would extend my sincere thanks to the chapter authors for their profound work. I am also thankful for the support provided by my family and colleagues during the compilation of this book.

Editor

Controlling Process of a Bottling Plant using PLC and SCADA

Kunal Chakraborty*[1], Indranil Roy[2], Palash De[3]
Department of Electrical Engineering, IMPS College of Engineering & Technology
Malda, West Bengal, 732103, India
e-mail: kunalindian003@gmail.com[1], indranilr6@gmail.com[2], palashde@hotmail.com[3]

Abstract

This paper presents basic stages of operation of a bottling plant, i.e. the filling and capping process. The main aim of our paper is to control the filling and capping section of a bottling plant simultaneously. At first a set of empty bottle is run by using a conveyer towards filling section, after the operation, the filled bottles are sent towards the capping section. After successful capping operation, the sealed bottles terminate towards exit and a new set of empty bottle arrive, in this way the process continues. This paper includes the method using which, a bunch of bottles can be filled and capped at one instant of time. This method has made the operation more flexible and time saving. The filling and capping operations are controlled using Programmable Logic Controllers (PLC), as the PLC's are very much user-efficient, cost-effective and easy to control. By using PLC automation the whole process is kept under control. SCADA (Supervisory Control and Data Acquisition) is used to monitor the process by means of a display system.

Keywords: PLC, automation, SCADA, ladder Logic, HMI

1. Introduction

Industrial Automation is the use of Control Systems to control Industrial Machinery and Processes, reducing the need for human intervention. If we compare a job being done by human and by Automation, the physical part of the job is replaced by use of a Machine, whereas the mental capabilities of the human are replaced with the Automation. The human sensory organs are replaced with electrical, mechanical or electronic Sensors to enable the Automation systems to perform the job.

Higher level of human intelligence like planning, analysis, prediction and intuitive decision making is not done by this Level of Automation.

Automation plays very important role in today's world economy. One of the most important applications of automation process is in beverages and soft drinks industries, where continuous filling and capping process is carried out. If human effort or mechanical effort is used in this field then it is very much tough to perform this long and continuous process and so it is being substituted by automation process which completes the task with very much ease.

As mentioned above, our paper is also an application where the automation process is used to control the filling and capping operation in a bottling plant to reduce the human effort using Programmable Logic Controllers and SCADA (Supervisory Control and Data Acquisition).

To develop the programming to control a bottling plant by using PLC Automation we must first develop the ladder logic, after that the programming part can be developed. After successful completion of the programming part, we have to animate the HUMAN- MACHINE INTERFACE or the HMI or SCADA.

2. Construction

The basic construction of the aforesaid processes of a bottling plant i.e. filling and capping is consisted of various steps. At first a conveyer belt is installed which will run the set of bottle through different stages. After that, in the filling section the necessary arrangements are done so that the filling process can take place by means of some filling pipes, containing the beverage or soft drink. In case of capping section also, some arrangements are done so that the capping process can be done without any error. To implement this steps, sensors are used so that in filling section, the pipes can sense the presence of the bottles and they can be filled. In capping section also, the sensors are used to cap the set of bottle with ease. The filling process is based on the preset value of a counter, depending upon which the pump is switched on for that particular period of time.

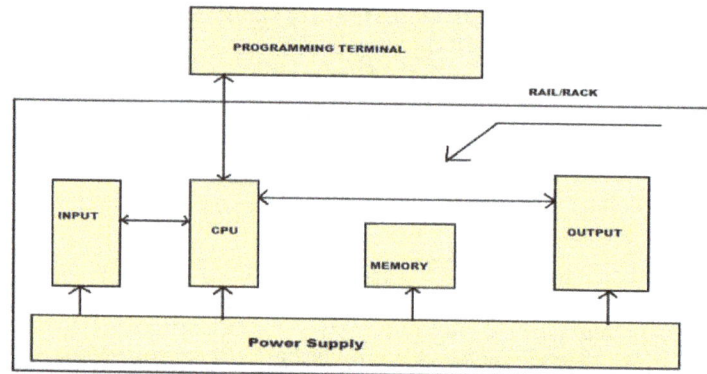

Figure 1. Rockwell Plc Structure

3. Process Description and Case Study in Factory

In our paper we have specialised on ROCKWELL PLC, in which the RSLOGIX 5000 software is the main platform to control the basic operations. The ladder logic, i.e. the programming part is done with the help of the above mentioned software. After successful completion of the programming, it is transferred to a virtual emulator which is already installed on the same workstation. As in case of a bottling plant, huge manpower is needed and as it is also very costly to implement the plant, we have given the basic priority to its security. The virtual emulator gives us the output whether the programming is correct or not. After the programming is made error free, it is installed on the main plc in the bottling plant.

Using plc programming the process of capping and filling is done simultaneously and as it is controlled by automation there is no need of constant manpower to handle the plant. There is one control room where the SCADA output is constantly observed by a person from where he can keep his close eye on various stages of the plant by using SCADA display. In case of emergency, the whole plant can be controlled from that control room only.

In this paragraph of our paper a detailed explanation of the various basic operations of a bottling plant is given. The filling and capping processes take place simultaneously.

At first an empty set of bottles are placed on a conveyer belt. When the conveyer is started, the empty set of bottles starts moving towards the filling section. After reaching the filling section the conveyer is stopped and the filling pipes then start filling the empty bottles. When the bottle filling is done then the conveyer again gets motion and the filled set of bottles move towards capping section. The set of bottles when reach the capping section again the conveyer gets stop and then capping process takes place. Completion of the capping process brings the conveyer again into motion and the set of filled and capped bottles move towards exit for further modification. This is a simultaneous process which is totally handled by PLC programming and in this way continuous filling and capping process takes place in a bottling plant.

Figure 2. Coca Cola Bottling Plant [11]

4. Control Philosophy

a) In a bottling plant there are two sections in it, Filling and Capping.
b) For the operation of the plant there will be 3 push-buttons.
c) The push-buttons will represent START, STOP, PAUSE.
d) The proximity sensor will sense the finished bottles as it passes by it in the conveyer belt.
e) The START button will start the whole system and also reset the counter to zero.
f) The STOP button will stop the whole system but it won't reset the counter value to zero, the numeric display will show the last counted value.
g) The RESET-COUNTER will reset the counter to zero.
h) If the PAUSE button is pressed then the system will hold its position and stop, and when it   is pressed again the system will resume.
i) Further modification can be done, i.e. a numeric display can be implemented through which the number of filled bottles can be monitored.

5. Ladder Logic

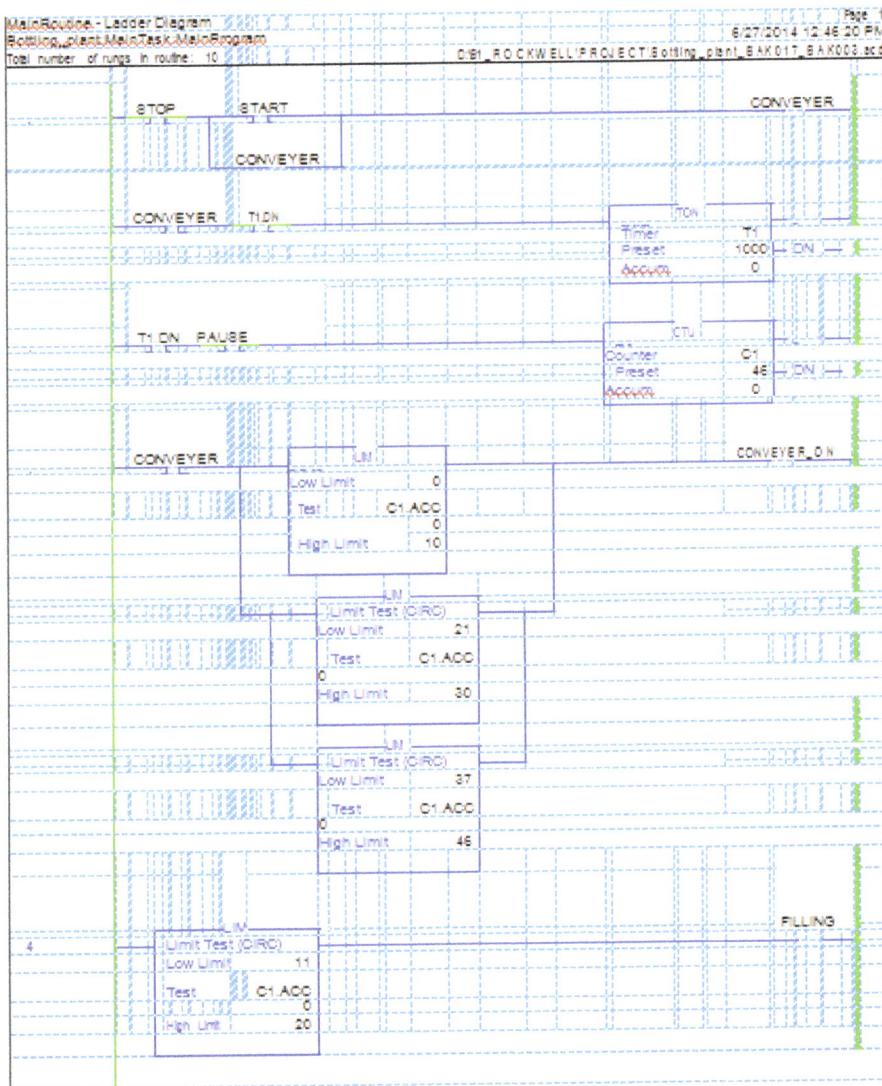

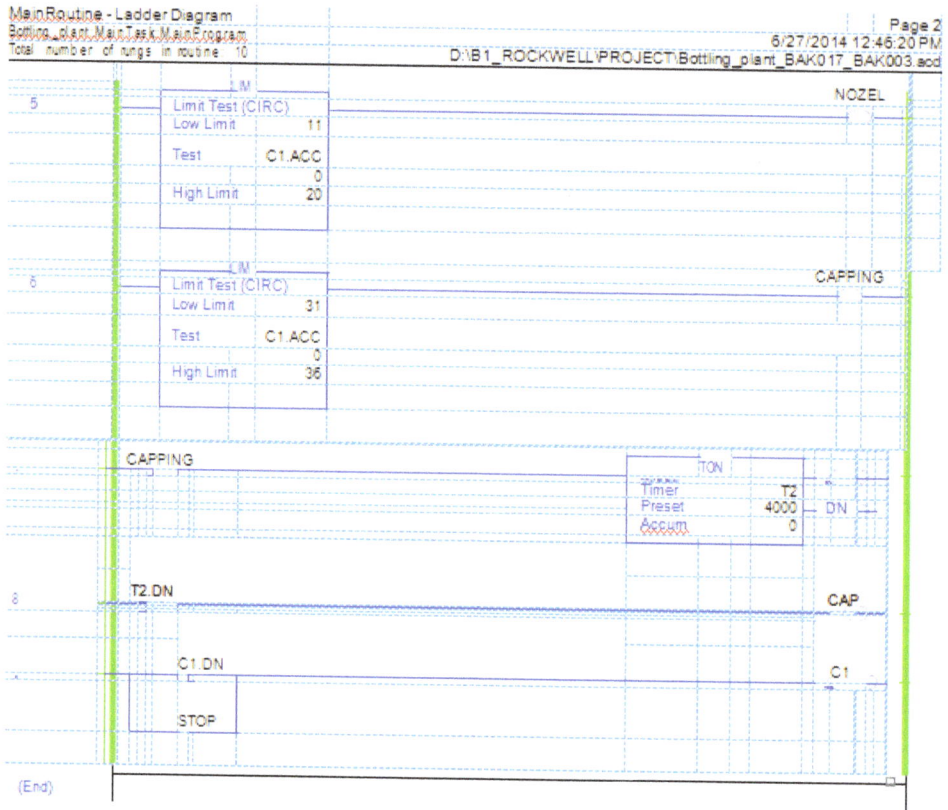

Figure 3. Ladder Diagram

6. SCADA Design

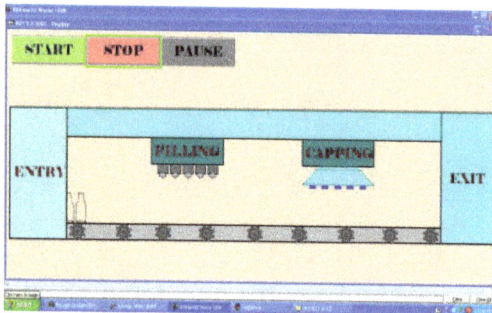

Figure 4(a). Conveyer On

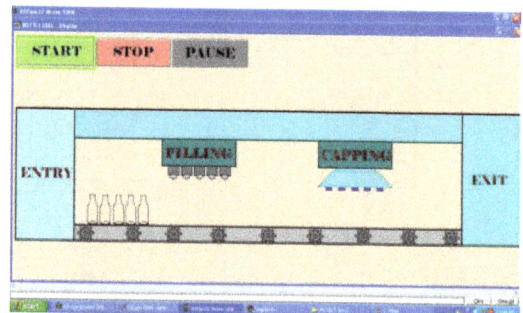

Figure 4(b). Empty Bottles Running

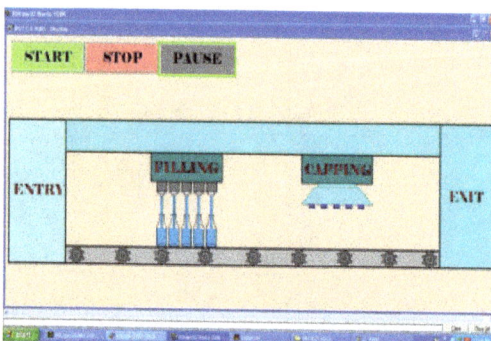

Figure 4(c). Bottles Filling

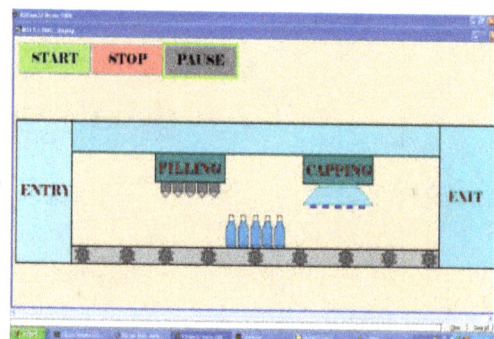

Figure 4(d). Filled Bottles Running

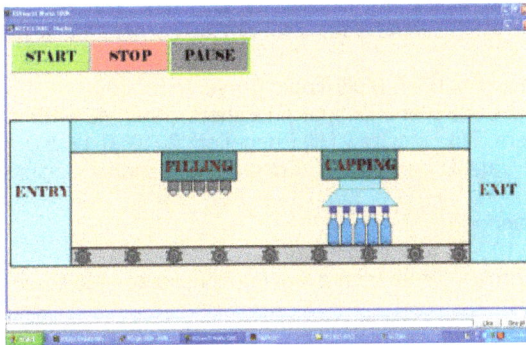

Figure 4(e). Capping Section   Figure 4(f). Capped Bottles Running

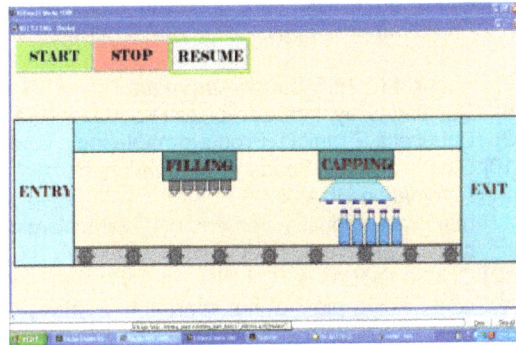

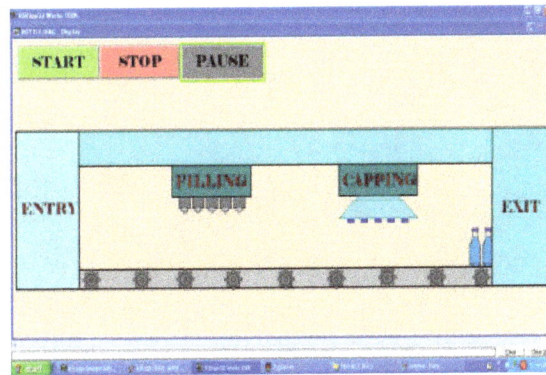

Figure 4(g). Set of Bottles Running Towards Exit

7. Conclusion

This paper has suggested the application of fully automated untouched plc controlled filling and capping operation of a bottling plant. The system works in high speed of production with very much accuracy and precision. This system meets the market demand with a few mechanical effort. The system has been proved working without wastage or spill out of the liquid. It is true that for small scale industries the installation cost ohf PLC is very much high but it has many advantages which overcomes the instllation cost. In this paper it is suggested how a set of bottle can be filled and capped at the same time. The other additional feature of this paper, here it is explained the SCADA design also. By using the SCADA the whole process can be monitored from a single control room only and necessary steps can be taken in case of emergency.

References

[1] T Kalaiselvi, R Praveena, Aakanksha R, Dhanya S. PLC Based Automatic Bottle Filling and Capping System with User Defined Volume Selection. *International Journal of Emerging Technology and Advanced Engineering*. 2(8).

[2] Ashwini P Somavanshi, Supriya B Asutkar, Sachin A More. Automatic Bottle Filling Using Microcontroller Volume Correction. *International Journal of Engineering Research and Technology IJERT*. 2013; 2(3): 1-4.

[3] Santhosh K V, JS Rajshekar. Design and Development of an Automated Multi Axis Solar Tracker Using PLC. *Bulletin of Electrical Engineering and Informatics*. 2013; 2(3).

[4] L Venkatesan, S Kanagavalli, PR Aarthi, KS Yamuna, PLC Scada Based Fault Identification and Protection for Three Phase Induction Motor. *TELKOMNIKA Indonesian Journal of Electrical Engineering*. 2014; 12(8).

[5] Ahmed Ullah Abu Saeed, Md Al-Mamun, AHM Zadidul Karim. Industrial Application of PLCs in Bangladesh. *International Journal of Scientific & Engineering Research*. 2012; 3(6).

[6] Mallaradhya HM, KR Prakash. *Automatic Liquid Filling to Bottles of Different Height Using Programmable Logic Controller*. In proceedings of AECE-IRAJ International Conference. 2013; 122-124.

[7] Shaukat N. *PLC Based Automatic Liquid Filling Process*. IEEE Multi Topic Conference. 2002.

[8] Dunning Gray. Introduction to Programmable Logic Controllers. Delmar publishers; 1998: 421-428.

[9] Petruzella, Frank D. Programmable logic Controllers. Tata McGraw Hill Education; 2010: 6-12.

[10] Stuart A Boyer. Scada – Supervisory Control and Data Acquisition. 4th Edition. International Society of Automation USA. 2009.

[11] http://www.industry.siemens.com/verticals/global/en/food-beverage/beverageindustry/Documents/E20001-A100-T110-V1-7600.pdf

[12] COCA COLA BOTTLING PLANT; http://www.profibus.com/technology/profibus/case-studies/coca-cola-bottling-plant-at-hm-interdrink-germany/

[13] www.rockwellautomation.com

2

Influence of Doping and Annealing on Structural, Optical and Electrical properties Amorphous ZnO Thin Films Prepared by PLD

Azhar AbduAlwahab Ali, K.T. Al-Rasoul, Issam M. Ibrahim
Iraqi Ministry of Sciences and Technology, Iraq

Abstract

In this work, ZnO thin films pure and doped with GaO were deposited using pulsed laser deposition (PLD). Technique using a double frequency Q-switching Nd:YAG laser beam (λ = 532) nm, repetition rate 6 Hz and the pulse duration 10 ns. After the end of wet mixing and drying process, the ZnO:GaO mixtures were pressed to form pellets (1.3 cm) diameter by using 3 ton pressure and sintered at 1373 K for 5 h. The product was investigated using XRD. The data of X-ray diffraction shows polycrystalline structure, and exhibited hexagonal structure. The film thickness was equal to 300 nm with rate of deposition of 0.5 nm/s. ZnO thin films pure and GaO-doped from the pellets with 0.02, 0.06 and 0.1 wt% of were deposited on glass substrates at room temperature. These films were annealed at different temperatures (373, 473 and 673K). The structural characteristics of the pure and GaO-doped ZnO films show that all the films have amorphous structure at room temperature and 373K, but when the samples are annealed at 473 and 673K; the XRD detected a hexagonal phase of ZnO. The surface morphology of the deposit materials was studied using atomic force microscope (AFM). The grain size of the particles observed at the surface depended on the annealing temperature. UV-VIS transmittance measurements showed that the films are highly transparent in the visible wavelength region for samples annealed up to 473K, while at annealing temperature of 673 K the absorption edge of ZnO doped with GaO was shifted to near-infrared region. The optical gap of the films was calculated from the curve of absorption coefficient $(\alpha h\upsilon)^2$ vs. $h\upsilon$ and was found to be 3.8 eV at room temperature, and this value decreases from 3.8 to 3.58 eV with increasing of annealing temperature up to 473-673 K, and increases with the Ga doping. λ_{cutoff} was calculated for ZnO and showed an increase with increasing annealing temperature and shifting to longer wavelength, while with doping the λ_{cutoff} shifted to shorter wavelength. The photoluminescence (PL) results indicate that the pure ZnO thin films grown at room temperature show strong peaks at 640 nm, but GaO doped ZnO films showed a band emission in the yellow-green spectral region (380 to 450nm).

Keywords: thin films, GaO doped ZnO, PLD, PL, electrical conductivity and resistivity

1. Introduction

Transparent conducting oxide (TCO) is very important in optoelectronic application, such as solar cell, sensor, and liquid crystal displays. In recent years, the appropriate materials of TCO are SnO_2, In_2O_3, $Sn:In_2O_3$ (ITO), Cd_2SnO_4 [1], and ZnO [2]. ITO is used usually to be a transparent conducting film [3-5], but the cost of ITO is too high to reduce the price of products which have a TCO film. ZnO is a semiconductor with a wide direct band gap (3.37 eV) and large exciton binding energy (60 meV). Exciton lasing mechanism from ZnO films at room temperature was reported recently [6]. Strong room temperature luminescence, high electron mobility, good transparency, etc. are some the advantages of ZnO [7, 8]. Wurtzite structured ZnO, a wide band gap semiconductor is a potential candidate for optoelectronics devices [9]. The conductivity of ZnO without intentionally doping is not high enough as TCO films. Thus, improving the conductivity of ZnO must rely upon doping elements into ZnO. The group-III elements, such as Al [10-14], Ga [15, 16], and In [17], are usually served as dopants for substituting zinc in order to increase more electronic carriers, then the conductivity can be improved. As doping concentration increased heavily, the amount of electronic carriers are also increased in general. Generally, ZnO films are fabricated by RF magnetron sputtering [18, 19], chemical vapor deposition [20], spray pyrolysis [21, 22] and sol–gel process [23, 24], etc. Among them, pulsed laser deposition [25] technique, metal-organic chemical vapour deposition (MOCVD) [26], arc plasma evaporation [27], dip-coating [28] and ion plating [29]. This outline provides a good context in which is pulsed laser deposition (PLD) can be viewed. PLD is a physical deposition technique: a physical process is used to deposit a vaporized form of the material onto a surface (substrate). No chemical reactions are involved. The advantages of pulsed laser ablation are flexibility, fast response, energetic evaporants and congruent evaporation

2. Experimentation

ZnO: GaO powders were mixed for 2h to obtained highly homogeneity samples. The powder then pressed to form a pellet of 1.3cm diameter at a pressure of 3 ton, using uniaxial hydraulic press. These pellets were then sintered at 1373K for 5hr.The concentrations of added oxide are given in Table 1.

Table 1. The concentration of added oxide

material ZnO	Doped with GaO
(9.998 gm)	0.002gm
(9.994 gm)	0.006gm
(9.990 gm)	0.010gm

3. Results and discussion

Figure 1 shows the X-ray diffraction pattern of ZnO thin films prepared by pulsed laser deposition (PLD) technique on glass substrate at room temperature with different annealing temperatures (373, 473K and 673K). There is not evidence of any phase present that means the formation of ZnO phase weak or amorphous. After annealing–at 673K for (2hr), X-ray diffraction detected a growth of ZnO on glass. We can be noticed from the X-ray pattern that the peaks at (2θ = 31.826 °, 34.481 °, 36.307 °, 56.598 °) referred to (100), (002), (101) and (110) crystalline planes, respectively. The X-ray diffraction data of thin films coincides with that of the known hexagonal structure. Table (2) shows the experiment and the standard peaks from International Centre for Diffraction Data **[Card No. (# 96-901-1663)]** of ZnO thin film annealed at 473K and 673K. The grain sizes of the prepared films after annealing 473K and 673K were calculated using the Scherrer's formula [30]:

$$D = k \lambda / \beta \cos\theta \qquad (1)$$

D:(G.S) is the grain size,
K: is a constant (0.94)
λ: is the wavelength of Cu Kα =1.54060 (^{0}A)
θ: is the Bragg's angle
β : Full Width at Half Maximum (FWHM)of the preferential plane.

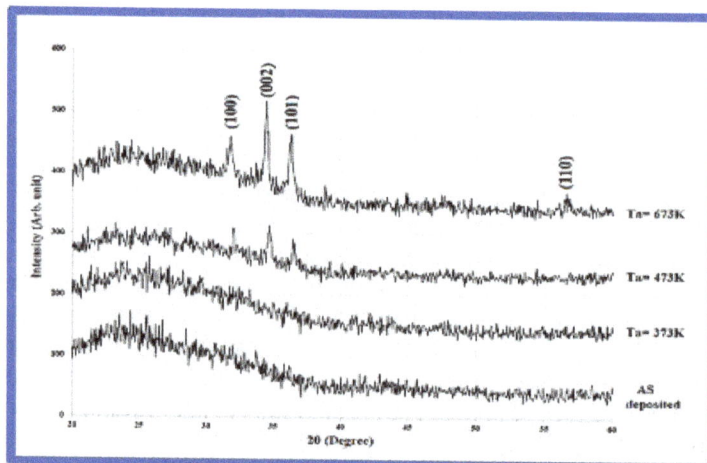

Figure 1. The X-ray diffraction patterns (XRD) of the un-doped ZnO thin film with different annealing temperature: (a) RT, (b) 373K, (c) 437 K, (d) 673K

Table 2. Represent the XRD parameters 2θ, hkl ,d exp, FWHM and Grain size for doped ZnO

\Ta (K)	2θ (deg)	hkl	FWHM (deg)	Int (arb. unit)	d_{hklExp}(Å)	d_{hklStd} (Å)	G.S (Å)	uniform strain*10^{-4}
303	-	-	-	-	-	-	-	-
373	-	-	-	-	-	-	-	-
473	32.033	(100)	0.332	52.917	2.7918	2.8137	235	-77.83
	34.730	(002)	0.332	70.828	2.5809	2.6035	236	-86.81
	36.432	(101)	0.332	55.360	2.4642	2.4754	237	-45.25
673	31.826	(100)	0.539	76.526	2.8095	2.8137	144	-14.93
	34.481	(002)	0.290	154.681	2.5990	2.6035	270	-17.28
	36.307	(101)	0.332	110.719	2.4724	2.4754	237	-12.12
	56.598	(110)	0.456	35.821	1.6249	1.6245	186	2.46

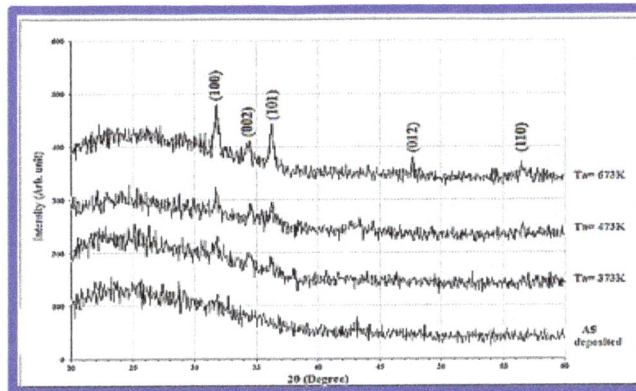

Figure 2. The X-ray diffraction patterns (XRD) of doped ZnO: Ga thin film with different annealing temperature: (a) RT, (b) 373K, (c) 437 K, (d) 673K

Table 3. Represent the XRD parameters 2θ, hkl ,d exp, FWHM and Grain size for Ga doped ZnO

Ta (K)	2θ (deg)	hkl	FWHM (deg)	Int (arb. unit)	d_{hklExp}(Å)	d_{hklStd} (Å)	G.S (Å)	uniform strain*10^{-4}
303	-	-	-	-	-	-	-	-
373	-	-	-	-	-	-	-	-
473	31.743	(100)	0.581	61.058	2.8167	2.8137	134	10.66
	34.564	(002)	0.373	38.263	2.5929	2.6035	210	-40.71
	36.224	(101)	0.415	52.103	2.4778	2.4754	190	9.70
	56.598	(110)	0.332	32.564	1.6249	1.6245	256	2.46
673	31.784	(100)	0.705	112.347	2.8131	2.8137	110	-2.13
	34.440	(002)	0.498	51.289	2.6020	2.6035	157	-5.76
	36.266	(101)	0.456	90.366	2.4751	2.4754	173	-1.21
	47.676	(012)	0.290	47.218	1.9060	1.9110	282	-26.16
	56.473	(110)	0.539	38.263	1.6282	1.6245	157	22.78

Figure 2 shows the XRD pattern for ZnO thin films doped with GaO in θ range from (20-60°), prepared by Pulsed laser deposition (PLD) technique on glass at room temperatures and with different annealing temperatures (373K, 473K and 673K). No evidence of any phases present on glass substrate as deposited that means the formation of ZnO phase is weak or amorphous. On the other hand, the XRD patterns of doped ZnO films with annealing temperature (673K) the structure of these films showed be a polycrystalline. We can be noticed from the X-ray pattern that the peaks at $(2\theta=31.784°, 34.440°, 36.266°, 47.676°$ and $56.473°)$ referred to (100), (002), (101), (012) and (110) direction, respectively. Table 3 shows experimental (2θ) shifting for (GaO) and (hkl) for film deposited on glass substrate.

3.2. Optical Microscopic Examination

Figure 3 shows the results of microscopic examination (Nikon- Japan) of the ZnO thin films before and after doping with GaO. Observed before doping the presence of two phases one crystalline of ZnO, and the other for amorphous phase as proved by XRD. In case of doping samples the surface appear clear that mean has less voids and looks smooth and homogenous.

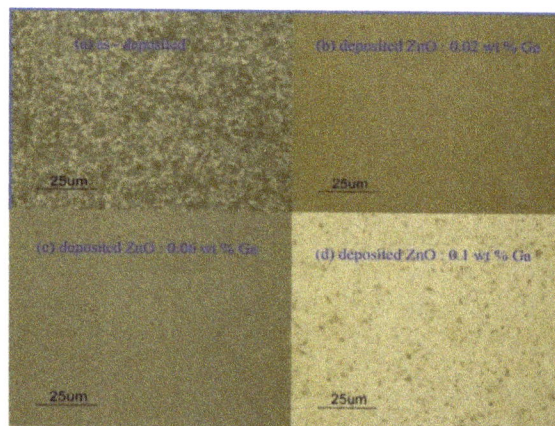

Figure 3. The image of Optical Microscopic of ZnO thin film. (a) ZnO without doped, (b) Doped with 2% Ga, (c) Doped with 0.06 wt% Ga, (d) Doped with0.1 wt % Ga

After annealing as show in Figure 4 and 5 observe the effect of annealing on the structural properties, when annealing at 473K note smoothing and clearly on the surface membranes, some disappearance of the voids and granular border as well as starts to crystallize and impurities are almost virtually non-existent, while annealing at 673K observe the disappearance of the most defects crystal line's improved the crystal structure and the surface becomes more homogeneity and fine.

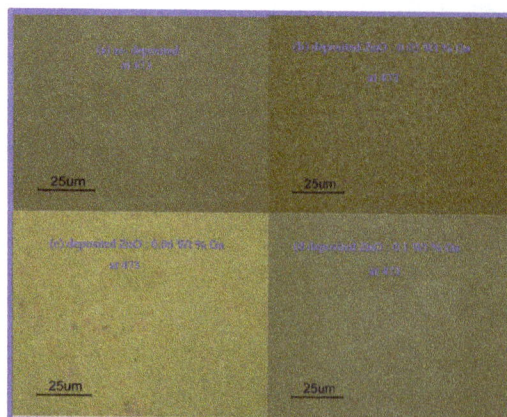

Figure 4. The image of Optical Microscopic of ZnO thin film annealing at 473K. (a) ZnO without doped, (b) Doped with 0.02 wt % Ga, (c) Doped with 0.06 wt% Ga, (d) Doped with 0.1 wt % Ga

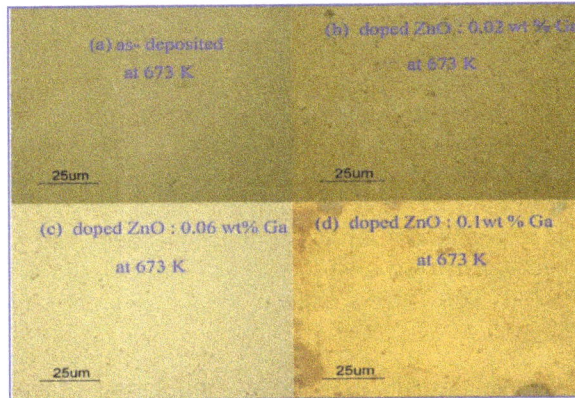

Figure 5. The image of Optical Microscopic of ZnO thin film annealing at 673K. (a) ZnO without doped, (b) Doped with 0.02 wt % Ga, (c) Doped with 0.06 wt% Ga, (d) Doped with 0.1 wt % Ga

3.3. Atomic Force Microscopy (AFM)

Figure 6 shows the influence of impurities on the grain size, where the doped leads to the decrease grain which that means with 0.06wt% Ga doping the crystalline nature of the film are decreased and also makes the surface more smoother and uniformed. In addition to taking into consideration the influence annealing at 673K on the grain size, the annealing cause increased the grain size and crystallized the membranes. Figure 7 shows the crystallization in the film was improved by a sufficient thermal crystallization at 673K.

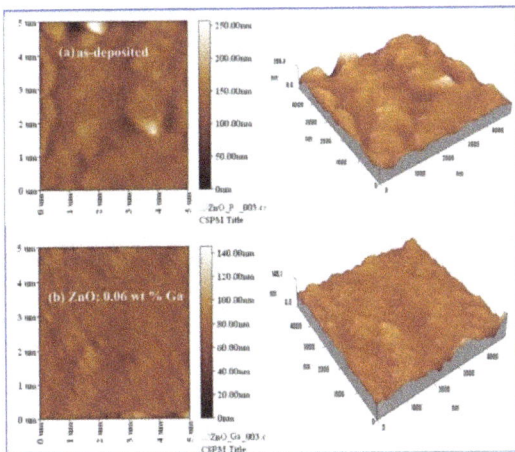

Figure 6. The AFM image of ZnO thin films at RT. (a) as- deposited, (b) Doped with 0.06 wt % Ga

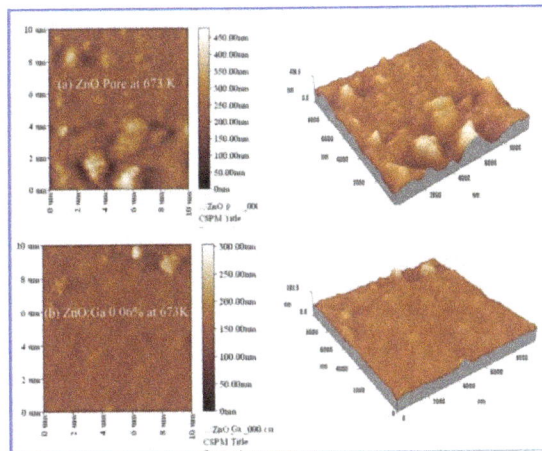

Figure 7. The AFM image of ZnO thin films annealing at 673K. (a) as- deposited, (b) Doped with 0.06 wt % Ga

5, Optical Properties of ZnO films

The optical properties of the deposited amorphous ZnO films on glass at room substrate temperature have thickness of (300) nm, at different annealing temperatures ranging from (373-673)K doping with different concentration of oxides, have been determined using UV-VIS in the region (200–1200) nm .The properties include the UV-VIS absorption, transmission spectrum have been measured. The optical energy gap is given by Tauc relationship [31].

$$\alpha h\upsilon = A(h\upsilon - E_g)^n \qquad (2)$$

Where, α is the absorption coefficient, A is the constant, h is the Planck's constant, υ is the photon frequency, Eg is the optical energy gap and n is the 1/2 for direct energy gap semiconductors. The optical energy gap decrease with increasing annealing temperature [32, 33], as show in Table 4. The direct energy gap values for amorphous ZnO pure and doped with different elements of mixed oxides (GaO) for doped is (0.02, 0.06 and 0.1)wt % are in the range of (3.8 –4.1) eV, as shown in the Figure 8. It is also observed that the direct energy gap energy inecreases with doping elements. This is presumably because of the effect exerted by the perturbation in the carrier concentration in the conduction band. The λ cut off calculate when wavelength = 0. The λ cut off increase shift to short of wavelength with increasing impurities ration [34] as show in Figure 7. The Table 5 explain effect doping on λ cut off.

Table 4. Gives the evolution of the band gap with deferent annealing temperature

Samples	Optical energy gab (eV) (direct)
ZnO at RT	3.8
ZnO at 373K	3.78
ZnO at473K	3.7
ZnO at 673K	3.58

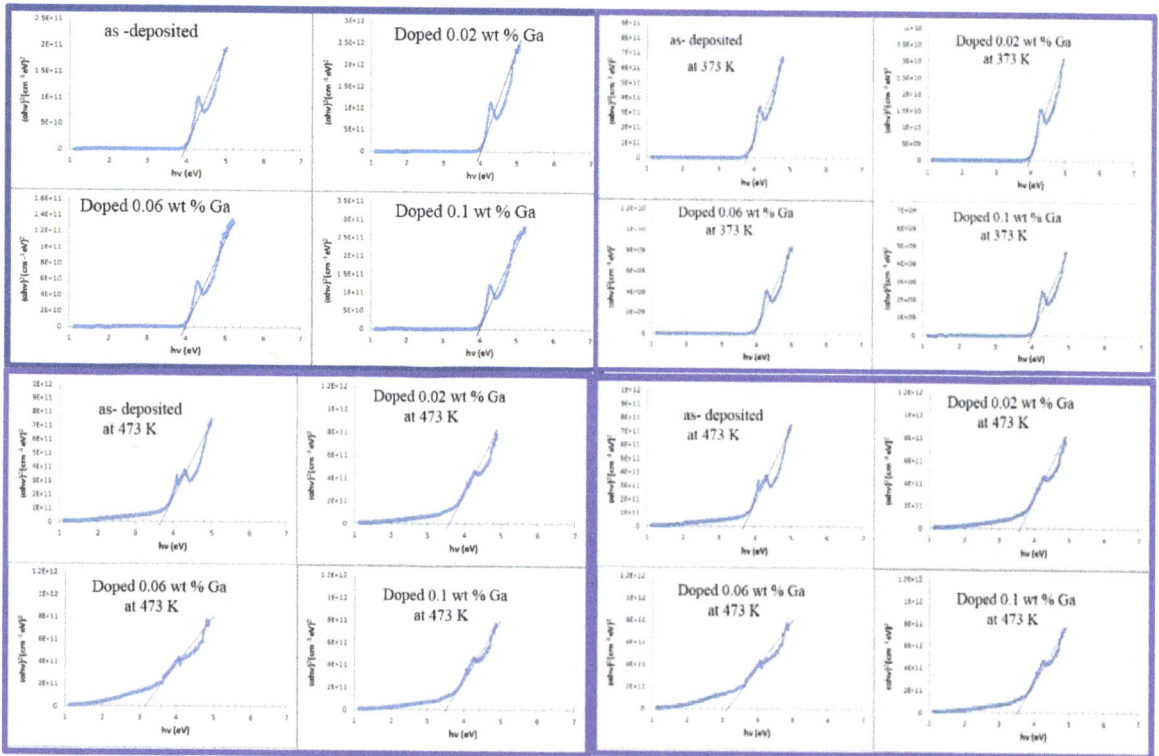

Figure 8. Measurement of energy band gap for ZnO pure and ZnO:Ga (0.02- 0.1%) thin films

Table 5. λ $_{cut off}$ values of ZnO undoped and doped thin film with deferent annealing temperature

Type of th film	λ $_{cut off}$ (nm) at RT	λ $_{cut off}$ (nm) at 373K	λ $_{cut off}$ (nm) at 473K	λ $_{cut off}$ (nm)at 673K
ZnO pure	340	350	350	358
ZnO:Ga0.02wt%	325	330	390	370
ZnO:Ga0.06wt%	320	325	395	380
ZnO:Ga0.1wt%	318	320	360	460

Figure 9 and 10 shows the spectral optical transmittance and absorbance as a function of wavelength in the range 200–1200 nm for amorphous ZnO thin films and doped with different elements of mixed oxides (GaO) of (0.02,0.06 and 0.1)wt⁒ on glass substrate at room temperature by pulse laser deposition. An increment in the transmittance is observed as the doping oxides were changed from (GaO). The films were found to be highly transmittance in the visible wavelength region. The maximum transmission observed for amorphous ZnO was almost (70%)up to 400nm, while for the doped films, the maximum transmittance equal (90%) for ZnO:GaO of (0.1 wt⁒).

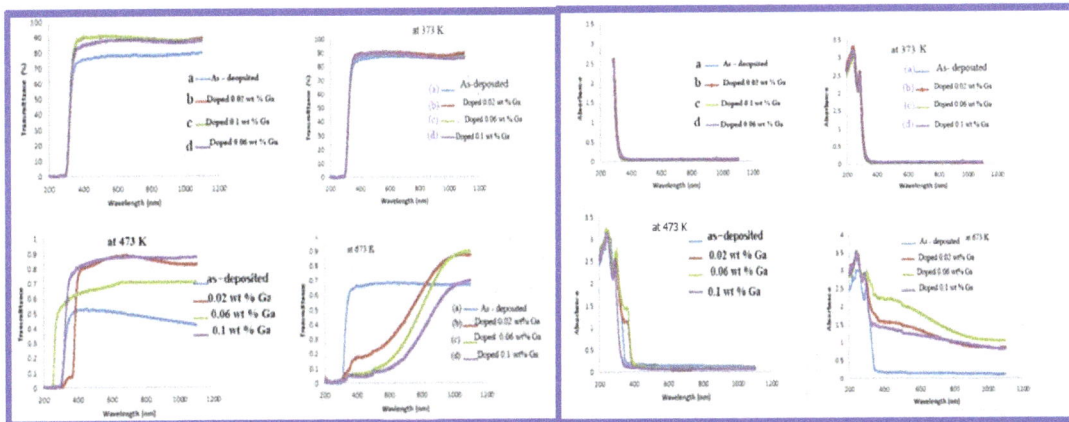

Figure 9. ZnO and ZnO: Ga (0.02-0.1) wt% thin films transmittance. a) ZnO pure, b) ZnO 0.02% Ga, c) ZnO 0.06 % Ga, d) ZnO 0.1% Ga with deferent tempareture

Figure10. ZnO and ZnO: Ga (0.02-0.1) wt% thin films absorbance. a) ZnO pure, b) ZnO 0.02% Ga, c) ZnO 0.06 % Ga, d) ZnO 0.1% Ga with deferent tempareture

The maximum transmittance observed for amorphous ZnO deposited at room temperature equal to (70%) in the UV region, while for the annealing films. the maximum transmittance equal (89%) at annealing temperature (373K). The behavior of the transmittance spectra is opposite completely to that of the absorption spectra. In general, we can observe from this figures that transmittance increases with increasing of annealing temperature and this may be due to improving the crystallite size which means a decrease in the absorption. The films were found to be highly transmittance in the visible wavelength region with an average transmittance in excess of 80%. This is probably ascribed to the increase of particle sizes and surface roughness.

The variation of absorption coefficient with wavelength for amorphous ZnO films deposited on glass substrate at room temperature at different annealing temperatures (373, 473 and 673) K are shown Figure 11. It is observed that the absorption coefficient decreases with increasing wavelength. This means that direct electronic transition happens. Also, we can notice from this figure that (α) in general increases with the incrase of annealing temperatures

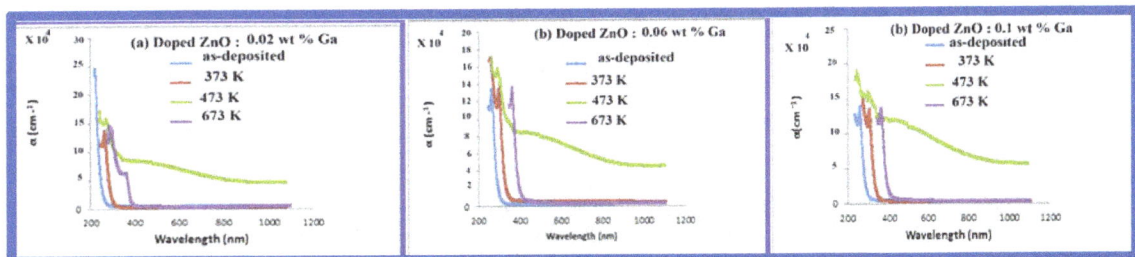

Figure 11. The absorption coefficient α (cm⁻¹) vs wavelength (nm) of ZnO undoped and doped thin films. a) ZnO 0.02% Ga, b) ZnO 0.06 % Ga, c) ZnO 0.1 % Ga calculate λ cut off values with different temperature

6. Conclusion

Undoped and doped ZnO with (Ga) oxide   films were successfully deposition on glass and silicon substrates by pulsed laser technique and thermally annealed at (373, 473 and 673)K, states were investigated. All films are semiconducting in nature with n-type conductivity.

The XRD results for undoped, doped and heated samples at 373K of ZnO thin films were amorphous. Thin films exhibit hexagonal crystal structure of undoped ZnO thin film at 473K and 673K. The intensity of peaks increased with increasing growth temperature at 673K.

The optical microscopic examination of thin film samples show the disappearance of voids, with smooth and relatively high homogeneity, AFM images also support the slow growth of crystallite sizes for the undoped and doped ZnO and also for the annealed films.

The transmission of undoped ZnO thin film was found to be above 70%, but higher for other doped one (85-90)%. After annealing an improvement were found in transmission for both undoped and doped film samples. In case of Samples annealed at 673K show low in transmission in UV region for ZnO thin film doped with Ga. The band gap for as prepaded was 3.8 eV ,then after annealed at 373K slightly decreased to be 3.78eV, then show a value of 3.70 eV at 473K and 3.59 eV at 673K. The $\lambda_{cut\ off}$ increase shift to shorter wavelength with doping, while the $\lambda_{cut\ off}$ increase with increasing annealing and shifted to longer wavelength.

PL emission spectrum from undoped ZnO thin film has a broad yellow-orange at wave length (610 nm). For doping ZnO with (0.02 and0.0 6)wt% Ga have a band emission in yellow-green regions (380-450)nm, In addition to the first emission(broad yellow-orange) and less intensity. ZnO doped with Ga 0.1wt%, give high intensity in the region of yellow-orange band, and then vanished at yellow-green band.

References

[1]   R Mamazza Jr, DL Morel, CS Ferekides. *Thin Solid Films*. 2005; 484(26).
[2]   DR Sahu, S Yuan, JL Huang. *Microelec*. 2007; 38(245).
[3]   GH Takaoka, D Yamazaki, J Matsuo. *Mat. Chem. Phys*. 2002; 74(104).
[4]   T Nakaoa, T Nakada, Y Nakayama, K Miyatania, Y Kimura, Y Saitob, C Kaitoa. *Thin Solid Films*. 2000; 307(155).
[5]   H Yumoto, T Inoue, SJ Li, T Sako, K Nishiyama. *Thin Solid Films*. 1999; 345(38).
[6]   ZK Tang, GKL Wong, P Yu, M Kawasaki, A Ohtomo, H Koinuma, Y Segawa. *Appl. Phys. Lett*. 1998; 72: 3270.
[7]   DC Look, DC Reynolds, JW Hemsky, RL Jones, JR Sizelove. *Appl. Phys. Lett*. 1999; 75: 811.
[8]   Y Caglar, S Ilican, M Caglar, F Yakuphanoglu. *Sol-Gel Sci Technol*. 2010; 53: 372.
[9]   KJ Kim, YR Park. *Appl Phys*. 2003; 94: 867.
[10] D Song, J Xia, AG Aberie. *Appl. Sur. Sci*. 2002; 195: 291.
[11] DC Altamirano-Juarez, G Torres-Delgado, S Jimenez-Sandoval, O Jimenez-Sandoval. *Sol. Ener. Mat. Sol. Cell*. 2004; 82: 35.
[12] RF Silva, MED Zaniquelli. *Thin Solid Films*, 2004; 449: 86.
[13] K Tominaga, T Murayama, I Mori, T Ushiro, T Moriga, I Nakabayashi. *Thin Solid Films*. 2001; 386: 267.
[14] MA Martinez, J Herrero, MT Gutierrez. *Sol. Ener. Mat. Sol. Cell*. 1997; 45: 75.
[15] JD Ye, SL Gu, SM Liu, YD Zheng, R Zhang, Y Shi. *Appl. Phys. Lett*. 2005; 86: 192111.
[16] JD Ye, SL Gu, S Zhu, SM Liu, YD Zheng, R Zhang, Y Shi, HQ Yu, YD Ye. *Cryst. Grow*. 2005; 283: 279.
[17] MT Young, SD Keun. *Thin Solid Films*. 2002; 410: 8.
[18] Y Zhou, PJ Kelly, A Postill, O Abu-Zeid, AA Alnajjar. *Thin Solid Films*. 2004; 33: 447–448.
[19] EG Fu, DM Zhuang, G Zhang, M Zhao, WF Yang. *Microelectr*. 2004; 35: 383.
[20] TM Barnes, J Leaf, C Fry, CA Wolden. *Cryst. Growth*. 2005; 274: 412.
[21] H Kim, A Pique, JS Horwitz, H Murata, ZH Kafafi, CM Gilmore, DB Chrisey. *Thin Solid Films*. 2000; 798: 377–378.
[22] SH Mondragon, A Maldonado, A Reyes. *Appl. Surf. Sci*. 2002; 193: 52.
[23] JH Lee, B O Park, *Mater. Sci. Eng. B*. 2004; 106: 242.
[24] Y Natsume, H Sakata. *Thin Solid Films*. 2000; 372: 30.
[25] GG Valle, P Hammer, SH Pulcinelli, CV Santilli. *Eur. Ceram. Soc*. 2004; 24: 1009.
[26] Kaul AR, Gorbenko OY, Botev AN, Burova LI. *Superlatt. Microstruct*. 2005; 38: 272.
[27] Minami T, Ida S, Miyata T, Minamino Y. *Thin Solid Films*. 2003; 445: 268.
[28] Fathollahi V, Amini MM. *Mater. Lett*. 2001; 50: 235.
[29] Iwata K, Sakemi T, Yamada A, Fons P, Awai K, Yamamoto T, Matsubara M, Tampo H, Niki S. *Thin Solid Films*. 2003; 445: 274.
[30] Th H DE Keijser, et al. *Appl. Cryst*. 1982; 15: 308-314.

[31] JI Pankove. Optical Processes in Semiconductors. Prentice-Hall, New Jersey. 1971.

[32] N Tigau, et al. *Optoelectronics Adv. Mater.* 2004; 6: 449.

[33] M Popa, V Lisca, M Stancu, E Buda, T Pentia. *Botila, Optoelectro. Adv. Mater.* 2006; 46.

[34] Hayder Mohammad Ajeel. Study effect of annealing Temperature on the structural and optical Properties of CdO Thin Films Prepared by SILAR Deposition Technique.

3

Study Pulse Parameters versus Cavity Length for both Dispersion Regimes in FM Mode Locked

Bushra R.Mhdi*, Gaillan H.Abdullah, Nahla A.Aljabar, Basher R.Mhdi
Ministry of Science and Technology, Iraq, Bagdad
e-mail: boshera65m@yahoo.com

Abstract

To demonstrate the effect of changing cavity length for FM mode locked on pulse parameters and make comparison for both dispersion regime, a plot for each pulse parameter as Lr function are presented for normal and anomalous dispersion regimes. The analysis is based on the theoretical study and the results of numerical simulation using MATLAB. The effect of both normal and anomalous dispersion regimes on output pulses is investigat Fiber length effects on pulse parameters are investigated by driving the modulator into different values. A numerical solution for model equations using fourth-fifth order, Runge-Kutta method is performed through MATLAB 7.0 program. Fiber length effect on pulse parameters is investigated by driving the modulator into different values of lengths. Result shows that, the output pulse width from the FM mode locked equals to τ= 501ns anomalous regime and τ=518ns in normal regime.

Keywords: FM mode lock, cavity length, dispersion regimes

1. Introduction

Ultrashort optical pulses have great applications in fields such as ultrafast optics, optical fibre communication, optical measurement, micro-mechanism processing, and medical treatments [1]. Compared with mode-locked solid-state lasers, ultrashort pulse fibre lasers have advantages of compactness, greater stability and so on. So far, researchers have proposed soliton fibre lasers, [2] stretched pulse fibre lasers [3] and self similar fibre laser. [4-5] But the evolution mechanisms of these pulse fibre lasers fail in output pulse energy or complexness (as dispersion compensation components, such as prism pairs, grating pairs, chirp mirrors, and micro construction fibres, are used). Recently, a novel all-normal-dispersion (ANDi) mode-locked fibre laser, in which all components are of normal dispersion, has been proposed [6]. In this diagram, the energy of output pulses can be much higher, [7-10] and designs can be greatly simplified especially for Yb-doped mode-locked fibre lasers whose operating wavelength is in the normal dispersion domain of common fibers. Ultrashort fiber lasers offer additional benefits such as small physical dimensions, increased stability under environmental conditions, reduced thermal management and diffraction limited beam quality. In addition, fiber lasers can be easily engineered and are usually less expensive than their solid-state lasers counterparts. Particularly, Yb3+ doped fibers have a broad emission bandwidth, large saturation fluence and high optical to optical conversion efficiency. Therefore, they form an excellent gain medium for the generation and amplification of ultrashort optical pulses in the 1µm wavelength range [11].

2. Theoretical Concept

Depending on master equation Eq.(1) , and using the assumed pulse shapes for both dispersion regimes after modifying Ginzburg–Landau Equation , GLE , by adding TOD and mode-locker effects, the extended solution will be as in the following relations [12-15].

$$T_R \frac{\partial A}{\partial T} + \frac{1}{2}\left(\overline{\beta_2} + i\overline{g}\,T_2^2\right)L_R \frac{\partial^2 A}{\partial t^2} - \frac{\overline{\beta_3}}{6}L\frac{\partial^3 A}{\partial t^3} = \overline{\gamma L}|A|^2\,A + \frac{1}{2}(\overline{g} - \overline{\alpha})LA + M(A,t) \qquad (1)$$

In the following all the terms of Mode-Locking master equation will be identified.

2.1. Normal Regime

$$A(T,t) = a \left(\exp\left(-\frac{(\Phi-\xi)^2}{2\tau^2}\right)\right)^{1+iq} \times \exp(i\Omega(t-\xi) + iKT + i\varphi_Q \tag{2}$$

2.2. Anomalous Regime

$$A(T,t) = a \left(\mathrm{sech}\left(\frac{t-\xi}{\tau}\right)\right)^{1+iq} \times \exp(i\Omega(t-\xi) + iKT + i\varphi_Q \tag{3}$$

Where pulse parameters for both profiles are:
   (a) represents pulse amplitude,
   (ξ) Temporal shift,
   (τ) pulse width,
   (q) Chirp,
   (Ω) frequency shift,
   $iKT + i\varphi_Q$ Represent the phase and rarely is of physical interest in lasers producing picoseconds pulses, and will be ignored [12].
   The relations of pulse parameters with temporal pulse profile are as following [16, 17]:

$$E(T) = \int_{-\infty}^{+\infty} |A(T,t)|^2 \, dt \tag{4}$$

$$\xi(T) = \frac{1}{E} \int_{-\infty}^{+\infty} t|A(T,t)|^2 \, dt \tag{5}$$

$$\Omega(T) = \frac{t}{2E} \int_{-\infty}^{+\infty} [A \frac{dA}{dt} - A \frac{dA}{dt}]^2 \, dt \tag{6}$$

$$q(T) = \frac{t}{E} \int_{-\infty}^{+\infty} (t-\xi)[A \frac{dA}{dt} - A \frac{dA}{dt}]^2 \, dt \tag{7}$$

$$(T)^2 = \frac{2}{E} \int_{-\infty}^{+\infty} (t-\xi)|A(T,t)^2 \, dt \tag{8}$$

3. Results and Analysis

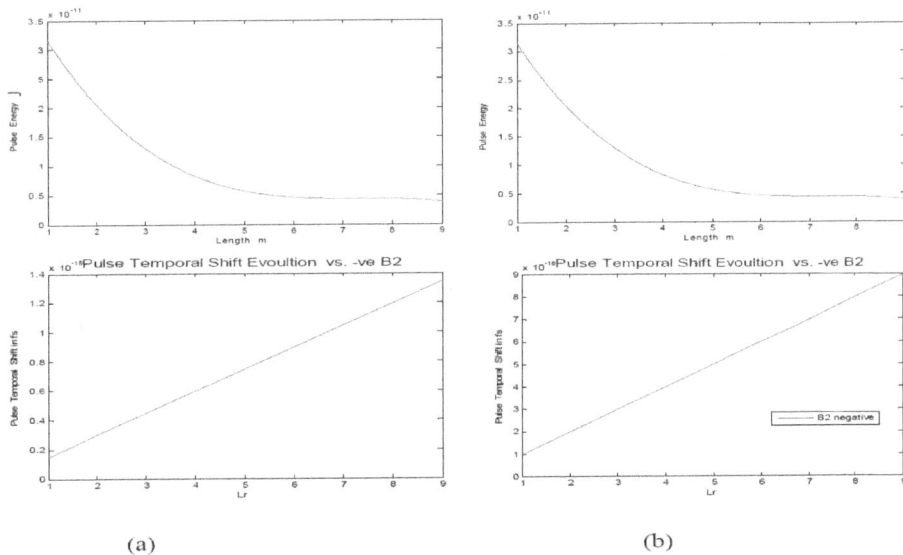

(a)   (b)

Figure 1. Pulse energy, temporal shift versus cavity length in (a) Normal, and (b) Anomalous dispersion

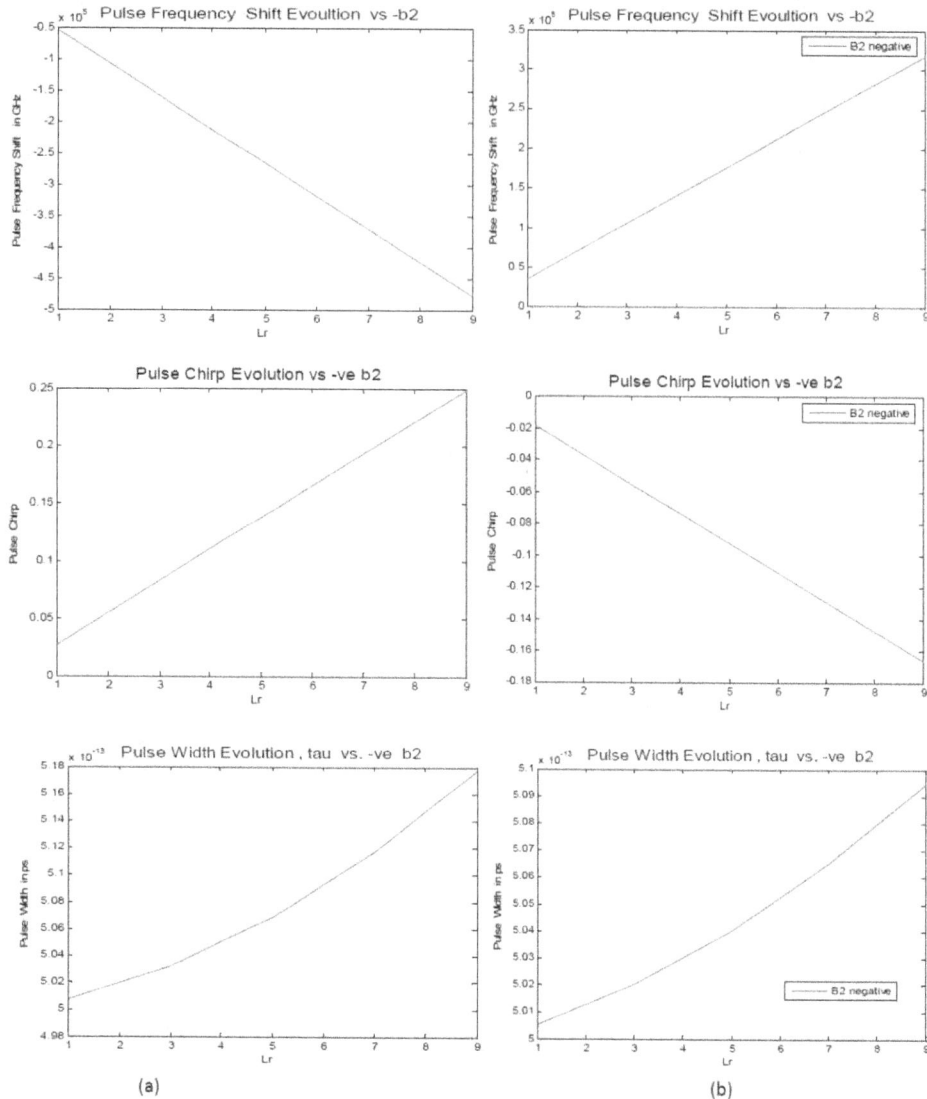

Figure 2. Frequency shift, chirp and width versus cavity length in (a) Normal, and(b) Anomalous dispersion

To demonstrate the effect of changing cavity length on pulse parameters and make comparison for both dispersion regimes, a plot for each pulse parameter as Lr function is drawn in Figure 1 for normal and anomalous dispersion regimes. As shown in Figure 1 almost same behavior for pulse energy is seen for variable Lr. Energy decreases exponentially as Lr increases. Then decreasing rate after certain value of cavity length (Lr = 5 m) and becomes so small for high values and almost constant straight line. Temporal shift plots in Figure 1 for both regimes are shown almost the same behavior in both regimes. Pulse temporal shift is increases linearly as Lr increases. The main difference is that: in anomalous regime, the temporal shift is greater than normal regime by five times. Inspecting plots for pulse frequency shift in Figure 2 the linear relation in anomalous and normal regime is seen. For normal regime, the frequency shift decrease linearly with Lr increases, while in anomalous regime, frequency shift increases with increase Lr.

For pulse chirp as function of cavity length, as shown in Figure 2 the pulse chirp in normal regime suffers from positive linear relation for variable Lr increasing with increasing of Lr .In anomalous regime, the negative linear relation chirp decreases with increasing Lr. This linear chirp enables us to compare the pulse outside the laser cavity without deforms in an anomalous dispersion regime until free chirp pulse is introduce [12]. Cavity Length has great effect on pulse width since it is responsible for mode-locking mechanism and producing pulses train. As shown in Figure 2, pulse width increases exponentially with the increase of Lr, but for anomalous regime it is much faster increasing in magnitude and less in Lr values as compared with normal regime. Almost, for anomalous regime, the minimum pulse width is obtained.

4. Conclusion

The length cavity varying of FM mode locking pulse parameter effect directly with varying Lr. The effective length in FM is equal 4m. Pulse width increases exponentially with the increase of Lr, but for anomalous regime it is much faster increasing in magnitude and less in Lr values as compared with normal regime. Almost, for anomalous regime, the minimum pulse width is obtained.

References

[1] U Keller, et al. Self-starting and reliable modelocking. *Opt. Lett.* 1992; 17: 505.
[2] V Matsas, T Newson, D Richardson, D Payne. vibrating soliton pairs in a mode- locked laser cavity. *Optics Letters.* 2006; 31(14): 2115-2117.
[3] H Haus, K Tamura, L Nelson, E Ippen.All normal – dispersion Yb-doped mode-locked fiber. *IEEE J. Quantum Electron.* 1995; 31: 591.
[4] FÖ Ilday, JR Buckley, WG Clark, FW Wise. All- fiber- integrated soliton- similiration laser with in line fiber filter. *Phys. Rev. Lett.* 2004; 92: 213902.
[5] Y Deng, Tu CH, Lu FY. All-normal dispersion Yb-doped mode-locked fiber. *Acta Phys Sin.* 2009; 58: 3173.
[6] A Chong, J Buckley, W Renninger, FW Wise. All-normal- dispersion femtosecond fiber laser. *Opt. Express.* 2006; 14: 10095.
[7] A Chong, W Renninger, F Wise. All-normal-dispersion femtosecond fiber laser with pulse energy above 20 nJ. *Opt. Lett.* 2007; 32: 2048.
[8] A Ruehl, V Kuhn, D Wandt, D Kracht. Generation of 1.7µJ pulses at 1,55µm. *Opt. Express.* 2008; 16: 3130.
[9] BW Liu, Hu ML, YJ Song, L Chai, QY Wang. Approaching 100nJ pulse energy output from a mode-locked. *Acta Phys. Sin.* 2008; 57: 6921.
[10] Y J Song, Hu M L, Q W Liu, J Y Li, W Chen, L Chai and Q Y Wang, "Dispersion- soliton mode-locked laser based on large-mode-area" *Acta Phys. Sin.* 57 , (2008).5045.
[11] A Chong, W Renninger, F Wise. Properties of normal-dispersion femtosecond fiber lasers. *Opt. Soc. Am.* 2008.
[12] G Nicholas Usechak, P Govind Agrawal, D Jonathan Zuegel. FM Mode-Locked Fiber Lasers Operating in the Autosoliton Regime. *IEEE Journal of Quantum Electronics.* 2005; 41(6).
[13] Xiangyu Zhou, Dai Yoshitomi, Yohei Kobayashi, Kenji Torizuka. Generation of 28-fs pulses from a mode-locked ytterbium fiber oscillator. *Optics Express.* 2008; 16(10): 2545-2505.
[14] G Nick Usechak, P Govind Agrawal, D Jonathan. Zuegel. Tunable high-repetition-rate, harmonically mode-locked ytterbium fiber laser. *Optics Letters.* 2004; 29(12): 2350-2365.
[15] J Jennifer ONeil. Pulse Dynamics in Actively Modelocking Fiber Optic Lasers. Honors Thesis Spring. University of Washington, Applied and Computational Mathematical Sciences. 2002.
[16] G Nicholas Usechak, P Govind Agrawal. Rate-equation approach for frequency-modulation mode locking using the moment method. *Opt. Soc.* 2005; 22(12).
[17] G Nicholas Usechak, P Govind Agrawal. Semi-analytic technique for analyzing mode-locked lasers. *Optics Express.* 2005; 13(6): 2075-2081.

4

Wolf Search Algorithm for Solving Optimal Reactive Power Dispatch Problem

K. Lenin*, B. Ravindhranath Reddy, M. Surya Kalavathi
Jawaharlal Nehru Technological University Kukatpally, Hyderabad 500 085, India
email: gklenin@gmail.com*

Abstract

This paper presents a new bio-inspired heuristic optimization algorithm called the Wolf Search Algorithm (WSA) for solving the multi-objective reactive power dispatch problem. Wolf Search algorithm is a new bio – inspired heuristic algorithm which based on wolf preying behaviour. The way wolves search for food and survive by avoiding their enemies has been imitated to formulate the algorithm for solving the reactive power dispatches. And the speciality of wolf is possessing both individual local searching ability and autonomous flocking movement and this special property has been utilized to formulate the search algorithm. The proposed (WSA) algorithm has been tested on standard IEEE 30 bus test system and simulation results shows clearly about the good performance of the proposed algorithm.

Keywords: *modal analysis, optimal reactive power, transmission loss, wolf search algorithm*

1. Introduction

Optimal reactive power dispatch problem is subject to number of uncertainties and at least in the best case to uncertainty parameters given in the demand and about the availability equivalent amount of shunt reactive power compensators. Optimal reactive power dispatch plays a major role for the operation of power systems, and it should be carried out in a proper manner, such that system reliability is not got affected. The main objective of the optimal reactive power dispatch is to maintain the level of voltage and reactive power flow within the specified limits under various operating conditions and network configurations. By utilizing a number of control tools such as switching of shunt reactive power sources, changing generator voltages or by adjusting transformer tap-settings the reactive power dispatch can be done. By doing optimal adjustment of these controls in different levels, the redistribution of the reactive power would minimize transmission losses. This procedure forms an optimal reactive power dispatch problem and it has a major influence on secure and economic operation of power systems. Various mathematical techniques like the gradient method [1, 2] Newton method [3]and linear programming [4-7] have been adopted to solve the optimal reactive power dispatch problem. Both the gradient and Newton methods has the difficulty in handling inequality constraints. If linear programming is applied then the input- output function has to be expressed as a set of linear functions which mostly lead to loss of accuracy. The problem of voltage stability and collapse play a major role in power system planning and operation [8]. Enhancing the voltage stability, voltage magnitudes within the limits alone will not be a reliable indicator to indicate that, how far an operating point is from the collapse point. The reactive power support and voltage problems are internally related to each other. This paper formulates by combining both the real power loss minimization and maximization of static voltage stability margin (SVSM) as the objectives. Global optimization has received extensive research attention, and a great number of methods have been applied to solve this problem. Evolutionary algorithms such as genetic algorithm have been already proposed to solve the reactive power flow problem [9, 10]. Evolutionary algorithm is a heuristic approach used for minimization problems by utilizing nonlinear and non-differentiable continuous space functions. In [11], by using Genetic algorithm optimal reactive power flow has been solved, and the main aspect considered is network security maximization. In [12] is proposed to improve the voltage stability index by using Hybrid differential evolution algorithm. In [13] Biogeography Based algorithm proposed to solve the reactive power dispatch problem. In [14] a fuzzy based method is used to solve the optimal reactive power scheduling method and it minimizes real power loss and maximizes Voltage Stability Margin. In [15] an improved evolutionary programming is used to solve the optimal reactive power dispatch problem. In [16] the optimal reactive power flow problem is solved by integrating a genetic algorithm with a nonlinear interior point method. In [17] a standard algorithm is used to solve ac-dc optimal reactive power flow model with the generator capability limits. In [18] proposed a two-step approach to evaluate Reactive power reserves with respect to operating constraints and voltage stability. In [19] a programming based proposed approach used to solve the optimal reactive power dispatch problem. In [20] is presented a probabilistic algorithm for optimal reactive power provision in hybrid electricity markets with uncertain loads. This research paper proposes a new bio-inspired heuristic search optimization algorithm, the Wolf Search Algorithm (WSA), for solving the optimal reactive power dispatch problem and this algorithm is based on wolf preying behaviour activity. Algorithm possesses both individual local searching ability and autonomous flocking movement. Wolf hunts independently by remembering its own trait and it will merge with its peer when the peer is in better position. The swarming behaviour of WSA has more advantage than that of algorithms like PSO [21], Fish [22] and Firefly [23]. WSA functions as multiple leaders swarming from multiple directions to reach the best solution, rather than searching as a single flock. How the wolf jumps far out of its hunter's visual range to avoid being trapped like that algorithm design will jump away from the local optimal solution. The wolves in the nature have best memory capability for they can hide food in caches; also they sense and track down a prey from distances of miles away. They themselves do set markers in their territory in various methods like by urinating at the borders. Researcher Sebastian Vetter and his team, from the University of Vienna have been studying the high level of observational spatial memory in the wolf. Main assumption is that the wolves are functioning as searching agents in the WSA optimization algorithm are empowered by memory caches that can able to store the previously visited various positions. The proposed algorithm WSA been evaluated in standard IEEE 30 bus test system & the simulation results shows that our proposed approach outperforms all reported algorithms in minimization of real power loss and voltage stability index .

2. Voltage Stability Evaluation

2.1. Modal Analysis for Voltage Stability Evaluation

Modal analysis is one among best methods for voltage stability enhancement in power systems. The steady state system power flow equations are given by.

$$\begin{bmatrix} \Delta P \\ \Delta Q \end{bmatrix} = \begin{bmatrix} J_{p\theta} & J_{pv} \\ J_{q\theta} & J_{QV} \end{bmatrix} \tag{1}$$

Where
ΔP = Incremental change in bus real power.
ΔQ = Incremental change in bus reactive
Power injection
$\Delta \theta$ = incremental change in bus voltage angle.
ΔV = Incremental change in bus voltage Magnitude

$J_{p\theta}$, J_{PV} , $J_{Q\theta}$, J_{QV} jacobian matrix are the sub-matrixes of the System voltage stability is affected by both P and Q.
To reduce (1), let $\Delta P = 0$, then:

$$\Delta Q = [J_{QV} - J_{Q\theta}J_{P\theta}{}^{-1}J_{PV}]\Delta V = J_R \Delta V \tag{2}$$

$$\Delta V = J^{-1} - \Delta Q \tag{3}$$

Where,

$$J_R = (J_{QV} - J_{Q\theta}J_{P\theta}{}^{-1}JPV) \tag{4}$$

J_R is called the reduced Jacobian matrix of the system.

2.2. Modes of Voltage Instability

Voltage Stability characteristics of the system have been identified by computing the Eigen values and Eigen vectors.

Let,

$$J_R = \xi \wedge \eta \tag{5}$$

Where,

ξ = right eigenvector matrix of J_R

η = left eigenvector matrix of J_R

$\wedge$ = diagonal eigenvalue matrix of J_R and

$$J_{R^{-1}} = \xi \wedge^{-1} \eta \tag{6}$$

From (5) and (8), we have:

$$\Delta V = \xi \wedge^{-1} \eta \Delta Q \tag{7}$$

Or,

$$\Delta V = \sum_I \frac{\xi_i \eta_i}{\lambda_i} \Delta Q \tag{8}$$

Where ξ_i is the ith column right eigenvector, and η is the ith row left eigenvector of JR. λ_i is the ith Eigen value of J_R.

The ith modal reactive power variation is:

$$\Delta Q_{mi} = K_i \xi_i \tag{9}$$

Where,

$$K_i = \sum_j \xi_{ij^2} - 1 \tag{10}$$

Where

ξ_{ji} is the jth element of ξ_i

The corresponding ith modal voltage variation is:

$$\Delta V_{mi} = [1/\lambda_i] \Delta Q_{mi} \tag{11}$$

If $| \lambda_i | = 0$ then the ith modal voltage will collapse.

In (10), let $\Delta Q = e_k$ where e_k has all its elements zero except the kth one being 1.

Then:

$$\Delta V = \sum_i \frac{\eta_{1k} \xi_1}{\lambda_1} \tag{12}$$

η_{1k} k th element of η_1

V –Q sensitivity at bus k

$$\frac{\partial v_K}{\partial Q_K} = \sum_i \frac{\eta_{1k} \xi_1}{\lambda_1} = \sum_i \frac{P_{ki}}{\lambda_1} \tag{13}$$

3. Problem Formulation

The objectives of the reactive power dispatch problem is to minimize the system real power loss and maximize the static voltage stability margins (SVSM).

3.1. Minimization of Real Power Loss

Minimization of the real power loss (Ploss) in transmission lines is mathematically stated as follows.

$$P_{loss} = \sum_{\substack{k=1 \\ k=(i,j)}}^{n} g_{k(V_i^2 + V_j^2 - 2V_i V_j \cos \theta_{ij})} \tag{14}$$

Where n is the number of transmission lines, gk is the conductance of branch k, Vi and Vj are voltage magnitude at bus i and bus j, and θij is the voltage angle difference between bus i and bus j.

3.2. Minimization of Voltage Deviation

Minimization of the voltage deviation magnitudes (VD) at load buses is mathematically stated as follows.

$$\text{Minimize VD} = \sum_{k=1}^{nl} |V_k - 1.0| \tag{15}$$

Where nl is the number of load busses and Vk is the voltage magnitude at bus k.

3.3. System Constraints

Objective functions are subjected to these constraints shown below.
Load flow equality constraints:

$$P_{Gi} - P_{Di} - V_i \sum_{j=1}^{nb} V_j \begin{bmatrix} G_{ij} & \cos \theta_{ij} \\ +B_{ij} & \sin \theta_{ij} \end{bmatrix} = 0, i = 1,2 \dots, nb \tag{16}$$

$$Q_{Gi} - Q_{Di} - V_i \sum_{j=1}^{nb} V_j \begin{bmatrix} G_{ij} & \cos \theta_{ij} \\ +B_{ij} & \sin \theta_{ij} \end{bmatrix} = 0, i = 1,2 \dots, nb \tag{17}$$

Where, nb is the number of buses, P_G and Q_G are the real and reactive power of the generator, P_D and Q_D are the real and reactive load of the generator, and G_{ij} and B_{ij} are the mutual conductance and susceptance between bus i and bus j.
Generator bus voltage (V_{Gi}) inequality constraint:

$$V_{Gi}^{min} \leq V_{Gi} \leq V_{Gi}^{max}, i \in ng \tag{18}$$

Load bus voltage (V_{Li}) inequality constraint:

$$V_{Li}^{min} \leq V_{Li} \leq V_{Li}^{max}, i \in nl \tag{19}$$

Switchable reactive power compensations (Q_{Ci}) inequality constraint:

$$Q_{Ci}^{min} \leq Q_{Ci} \leq Q_{Ci}^{max}, i \in nc \tag{20}$$

Reactive power generation (Q_{Gi}) inequality constraint:

$$Q_{Gi}^{min} \leq Q_{Gi} \leq Q_{Gi}^{max}, i \in ng \tag{21}$$

Transformers tap setting (T_i) inequality constraint:

$$T_i^{min} \leq T_i \leq T_i^{max}, i \in nt \tag{22}$$

Transmission line flow (S_{Li}) inequality constraint:

$$S_{Li}^{min} \leq S_{Li}^{max}, i \in nl \tag{23}$$

Where, nc, ng and nt are numbers of the switchable reactive power sources, generators and transformers.

4. Wolf Search Algorithm

Wolves are social predators that hunt in packs and uses stealth when hunting prey together. In behaviour of ants it utilizes pheromones to communicate with their peers to know about food source. WSA [24] also do this kind of communication, which decreases the run time of the search. Wolves are unique, partially cooperative characteristics and usually move in a group in coupled formation, but have tendency to take down the prey individually. WSA naturally balances scouting the problem space in random groups and individual. During hunting, wolves will group themselves as they approach their prey. This peculiar characteristic prompts the searching agents in WSA to move for a better position, like the same way wolves continuously change their positions for better ones. When hunting, wolves search for prey and also keenly watch the threats from hunters or other animals like tigers etc. Each wolf in the pack chooses its own way & position continuously moving to a better state for the prey and also for threats in all directions. When wolves' bumping into their enemies it is well equipped with a threat probability and it dashes a great distance away from its present position. The same way in WSA avoids the deadlock of getting trapped in local optimal solution. The direction and distance the wolf moving away from a threat are random, and is similar to mutation and crossover in Genetic algorithm .Wolves have very high sense of smell and it can easily locate prey by scent. Similarly, in the WSA each wolf has a sensing distance that creates visual distance. This visual distance is applied to search the global optimum and in moving to a better position and for jumping out of visual range. In search mode, the wolves are move in Brownian motion (BM), which imitates the random drifting of particles suspended in fluid.

Basic logics of wolf search

There are three rules that act as basic logics of the Wolf Search Algorithm (WSA)

Rule 1: Each wolf has visual area as a fixed one and with a radius defined by v for X as a set of continuous possible solutions. Each wolf can sense companions who are all appear within its visual circle. The footstep expanse by which the wolf moves at a time is normally smaller than its visual distance.

Rule 2: The fitness of the objective function represents the wolf's current position. If there is more options the wolf will chose the best terrain inhabited by another wolf from the given options. If not, the wolf will continue to move randomly in BM.

Rule 3: if the wolf will sense an enemy then the wolf will immediately escape to a random position far from the threat and beyond its visual range.

WSA implementation in based on the fitness of the objective function and it reflects the quality of a terrain position which will eventually lead to food.

Wolf often changes in position in search of food and also to safeguard form the enemies. Wolf trust with other wolves in movement because they never prey each other. The movement done by one wolf will be watched by other wolves simultaneously and they position themselves in chance of finding food also with care of them by continuously moving. If the current wolf's location is greater the distance of the companion location, then that new location will be less attractive one even though the new position may be good one. Wolf's willingness to move is decreased means, and then that movement will obey the inverse square law. The formula is $(r) = \frac{I_0}{r^2}$, where Io is the origin of food and r is the distance between the food or we can denote that distance between the new terrain and the wolf.

This is the similar formula in the firefly algorithm, for the calculation of attractiveness. The incentive formula for the wolf search by using absorption coffeicient and gaussian equation, can be written as:

$$\beta(r) = \beta_o e^{-r^2} \tag{24}$$

Normally all the wolves want to move better position based on colonized by their peers position and it depends on many factors like visual distance and how the initial wolf covers the area. Wolf will visualize the other wolves location each other i.e. it will compare the range of distance and set by itself in best position for preying and also from enemies. The movement can be written as:

$$x(i) = x(i) + \beta_o e^{-r^2}\big(x(j) - x(i)\big) + escape(\,) \tag{25}$$

Where, escape ( ) is a function that calculates a random position to jump to with a constraint of minimum length; v, x is the wolf, which represents a candidate solution; and x(j) is the peer with a better position as represented by the value of the fitness function. The second term of the above equation represents the change in value or gain achieved by progressing to the new position. r is the distance between the wolf and its peer with the better location.

There are three types of preying that takes place in sequence,

1) Preying initiatively

Wolf feed on prey it represents the optimization function as objective. By using the visual boundary wolf will have step by step movement on constantly seeing the prey and it will have random movement from the current step to forward or backward depending on the prey position. If it thinks particular position as best one then it will omit other wolves movements. Then it will move in own direction.

2) Prey passively

In passive mode the wolf will compare the position with its peers and will improve the current position. Wolf will move to passive mode when its own movement does not find food or insecurity for its movement.

3) Escape

Wolves normally have enemies in nature and threat will be there always. If any threat is found, it will relocate very quickly form the current position to new position which will be normally greater distance than that of the normal visual range. This can be written in equation as:

$$\text{if moving} = \begin{cases} x(i) = x(i) + \alpha \cdot r \cdot \text{rand}(\,)\text{prey} \\ x(i) = x(i) + \alpha \cdot s \cdot \text{escape} (\,)\text{escape} \end{cases} \tag{26}$$

Where x(i) is the wolf's location; a is the velocity; v is the visual distance; rand() is a random function whose mean value distributed in [-1,1], s is the step size, which must be smaller than v; and escape() is a custom function that randomly generates a position greater than v and less than half of the solution boundary.

Wolf algorithm for solving optimal reactive power dispatch problem:
Step 1: Objective function f(x), x =$(x_1,x_2,..xd)^T$
Step 2: Initialize the population, x_i(i=1,2,..,W)
Step 3: initialize parameters
r = radius of the visual range
s = step size by which a wolf moves at a time
α = velocity factor of wolf
pa = a user-defined threshold [0-1], determines how often foe appears
Step 4: WHILE (t<generations and also for stopping criteria is not met)
step5: FOR i=1: W // each wolf
step6: Prey new food initiatively ();
step7: Generation of new location ();
step8: To check whether the next location suggested by the random number generator is new one .
step8: If not, repeat generating random location.
Step9:IF(dist(x_i,x_j) < r and x_j is better as f(x_i)<f(x_j)) x_i moves towards x_j // x_j is a better than x_i
Step 10: ELSE IF
x_i = Prey new food passively ();
Step 11: END IF
Generation of new location ();
IF (rand ()>pa)
x_i = x_i + rand() + v;   wolf escape to a new position.
END IF
END FOR
END WHILE

5. Simulation Results

The accurateness of the proposed WSA method is demonstrated by testing it on standard IEEE-30 bus system. The IEEE-30 bus system has 6 generator buses, 24 load buses and 41 transmission lines of which four branches are (6-9), (6-10), (4-12) and (28-27) - are with the tap setting transformers. The lower voltage magnitude limits at all buses are 0.95 p.u. and the upper limits are 1.1 for all the PV buses and 1.05 p.u. for all the PQ buses and the reference bus. The simulation results have been presented in Tables 1, 2, 3 &4. And in the Table 5 shows the proposed algorithm powerfully reduces the real power losses when compared to other given algorithms. The optimal values of the control variables along with the minimum loss obtained are given in Table 1. Corresponding to this control variable setting, it was found that there are no limit violations in any of the state variables.

Table 1. Results of WSA – ORPD optimal control variables

Control variables	Variable setting
V1	1.041
V2	1.042
V5	1.041
V8	1.031
V11	1.002
V13	1.040
T11	1.00
T12	1.00
T15	1.02
T36	1.01
Qc10	3
Qc12	3
Qc15	4
Qc17	0
Qc20	3
Qc23	4
Qc24	4
Qc29	3
Real power loss	4.3209
SVSM	0.2462

Table 2. Results of WSA -Voltage Stability Control Reactive Power Dispatch Optimal Control Variables

Control Variables	Variable Setting
V1	1.043
V2	1.043
V5	1.042
V8	1.031
V11	1.006
V13	1.034
T11	0.090
T12	0.090
T15	0.090
T36	0.090
Qc10	4
Qc12	4
Qc15	3
Qc17	3
Qc20	0
Qc23	3
Qc24	2
Qc29	3
Real power loss	4.9870
SVSM	0.2471

ORPD together with voltage stability constraint problem was handled in this case as a multi-objective optimization problem where both power loss and maximum voltage stability margin of the system were optimized simultaneously. Table 2 indicates the optimal values of these control variables. Also it is found that there are no limit violations of the state variables. It indicates the voltage stability index has increased from 0.2462 to 0.2471, an advance in the system voltage stability. To determine the voltage security of the system, contingency analysis was conducted using the control variable setting obtained in case 1 and case 2. The Eigen values equivalents to the four critical contingencies are given in Table 3. From this result it is observed that the Eigen value has been improved considerably for all contingencies in the second case.

Table 3. Voltage Stability under Contingency State

Sl.No	Contingency	ORPD Setting	VSCRPD Setting
1	28-27	0.1410	0.1427
2	4-12	0.1658	0.1668
3	1-3	0.1774	0.1784
4	2-4	0.2032	0.2047

Table 4. Limit Violation Checking Of State Variables

State variables	limits Lower	upper	ORPD	VSCRPD
Q1	-20	152	1.3422	-1.3269
Q2	-20	61	8.9900	9.8232
Q5	-15	49.92	25.920	26.001
Q8	-10	63.52	38.8200	40.802
Q11	-15	42	2.9300	5.002
Q13	-15	48	8.1025	6.033
V3	0.95	1.05	1.0372	1.0392
V4	0.95	1.05	1.0307	1.0328
V6	0.95	1.05	1.0282	1.0298
V7	0.95	1.05	1.0101	1.0152
V9	0.95	1.05	1.0462	1.0412
V10	0.95	1.05	1.0482	1.0498
V12	0.95	1.05	1.0400	1.0466
V14	0.95	1.05	1.0474	1.0443
V15	0.95	1.05	1.0457	1.0413
V16	0.95	1.05	1.0426	1.0405
V17	0.95	1.05	1.0382	1.0396
V18	0.95	1.05	1.0392	1.0400
V19	0.95	1.05	1.0381	1.0394
V20	0.95	1.05	1.0112	1.0194
V21	0.95	1.05	1.0435	1.0243
V22	0.95	1.05	1.0448	1.0396
V23	0.95	1.05	1.0472	1.0372
V24	0.95	1.05	1.0484	1.0372
V25	0.95	1.05	1.0142	1.0192
V26	0.95	1.05	1.0494	1.0422
V27	0.95	1.05	1.0472	1.0452
V28	0.95	1.05	1.0243	1.0283
V29	0.95	1.05	1.0439	1.0419
V30	0.95	1.05	1.0418	1.0397

Table 5. Comparison of Real Power Loss

Method	Minimum loss
Evolutionary programming [25]	5.0159
Genetic algorithm [26]	4.665
Real coded GA with Lindex as SVSM [27]	4.568
Real coded genetic algorithm [28]	4.5015
Proposed WSA method	4.3209

6. Conclusion

In this paper, the WSA has been successfully implemented to solve optimal reactive power dispatch (ORPD) problem. The main advantages of WSA when applied to the ORPD problem is optimization of different type of objective function, i.e real coded of both continuous and discrete control variables, and without difficulty in handling nonlinear constraints. The proposed WSA algorithm has been tested on the IEEE 30-bus system. Simulation Results clearly show the good performance of the proposed algorithm in reducing the real power loss and enhancing the voltage profiles within the limits.

References

[1]   O Alsac, B Scott. Optimal load flow with steady state security. *IEEE Transaction.* 1973; 745-751.
[2]   Lee KY, Paru YM, Oritz JL. A united approach to optimal real and reactive power dispatch. *IEEE Transactions on power Apparatus and systems.* 1985; 104: 1147-1153
[3]   A Monticelli, MVF Pereira, S Granville. Security constrained optimal power flow with post contingency corrective rescheduling. *IEEE Transactions on Power Systems:PWRS-2.* 1987; 1: 175-182.
[4]   Deeb N, Shahidehpur SM. Linear reactive power optimization in a large power network using the decomposition approach. *IEEE Transactions on power system.* 1990; 5(2): 428-435.
[5]   E Hobson. Network consrained reactive power control using linear programming. *IEEE Transactions on power systems PAS -99.* 1980; (4): 868-877.
[6]   KY Lee, YM Park, JL Oritz. *Fuel –cost optimization for both real and reactive power dispatches.* IEE Proc. 131C(3): 85-93.
[7]   MK. Mangoli, KY Lee. Optimal real and reactive power control using linear programming. *Electr.Power Syst.Res.* 1993; 26: 1-10.
[8]   CA Canizares, ACZ de Souza, VH Quintana, Comparison of performance indices for detection of proximity to voltage collapse. 1996; 11(3): 1441-1450.
[9]   SR Paranjothi, K Anburaja. Optimal power flow using refined genetic algorithm. *Electr.Power Compon.Syst.* 2002; 30: 1055-1063.
[10] D Devaraj, B Yeganarayana. *Genetic algorithm based optimal power flow for security enhancement.* IEE proc-Generation Transmission and Distribution. 2005; 152.
[11] A Berizzi, C Bovo, M Merlo, M Delfanti. A ga approach to compare orpf objective functions including secondary voltage regulation. *Electric Power Systems Research.* 2012; 84(1): 187–194.
[12] CF Yang, GG Lai, CH Lee, CT Su, GW Chang. Optimal setting of reactive compensation devices with an improved voltage stability index for voltage stability enhancement. *International Journal of Electrical Power and Energy Systems.* 2012; 37(1): 50–57.
[13] P Roy, S Ghoshal, S Thakur. Optimal var control for improvements in voltage profiles and for real power loss minimization using biogeography based optimization. *International Journal of Electrical Power and Energy Systems.* 2012; 43(1): 830–838.
[14] B Venkatesh, G Sadasivam, M Khan. A new optimal reactive power scheduling method for loss minimization and voltage stability margin maximization using successive multi-objective fuzzy lp technique. *IEEE Transactions on Power Systems.* 2000; 15(2): 844 – 851.
[15] W Yan, S Lu, D Yu. A novel optimal reactive power dispatch method based on an improved hybrid evolutionary programming technique. *IEEE Transactions on Power Systems.* 2004; 19(2): 913 – 918.
[16] W Yan, F Liu, C Chung, K Wong. A hybrid genetic algorithminterior point method for optimal reactive power flow. *IEEE Transactions on Power Systems.* 2006; 21(3): 1163 –1169.
[17] J Yu, W Yan, W Li, C Chung, K Wong. An unfixed piecewiseoptimal reactive power-flow model and its algorithm for ac-dc systems. *IEEE Transactions on Power Systems.* 2008; 23(1): 170 –176.
[18] F Capitanescu. Assessing reactive power reserves with respect to operating constraints and voltage stability. *IEEE Transactions on Power Systems.* 2011; 26(4): 2224–2234.
[19] Z Hu, X Wang, G. Taylor. Stochastic optimal reactive power dispatch: Formulation and solution method. *International Journal of Electrical Power and Energy Systems.* 2010; 32(6): 615 – 621.
[20] A Kargarian, M Raoofat, M Mohammadi. Probabilistic reactive power procurement in hybrid electricity markets with uncertain loads. *Electric Power Systems Research.* 2012; 82(1): 68 – 80.
[21] XS Yang, S Deb, S Fong. *Accelerated Particle Swarm Optimization and Support Vector Machine for Business Optimization and Applications.* The Third International Conference on Networked Digital Technologies (NDT 2011), Springer CCIS 136. Macau. 2011; 53–66.
[22] Y Peng. An Improved Artificial Fish Swarm Algorithm for Optimal Operation of Cascade Reservoirs. *Journal of Computers.* 2011; 6(4): 740–746.
[23] XS Yang. Firefly algorithms for multimodal optimization. Stochastic Algorithms: Foundations and Applications. Lecture Notes in Computer Sciences. 2009; 169–178.
[24] Rui Tang, Simon Fong, Xin-She Yang, Suash Deb. *Wolf Search Algorithm with Ephemeral Memory.* Seventh International Conference on Digital Information Management, ICDIM 2012. Macao. 2012.
[25] Wu QH, Ma JT. Power system optimal reactive power dispatch using evolutionary programming. *IEEE Transactions on power systems.* 1995; 10(3): 1243-1248.
[26] S Durairaj, D Devaraj, PS Kannan. Genetic algorithm applications to optimal reactive power dispatch with voltage stability enhancement. *IE(I) Journal-EL.* 2006; 87.
[27] D Devaraj. Improved genetic algorithm for multi – objective reactive power dispatch problem. *European Transactions on electrical power.* 2007; 17: 569-581.
[28] P Aruna Jeyanthy, Dr D Devaraj. Optimal Reactive Power Dispatch for Voltage Stability Enhancement Using Real Coded Genetic Algorithm. *International Journal of Computer and Electrical Engineering.* 2010; 2(4): 1793-8163.

5

Modified Monkey Optimization Algorithm for Solving Optimal Reactive Power Dispatch Problem

K. Lenin, B. Ravindhranath Reddy, M. Suryakalavathi
Jawaharlal Nehru Technological University Kukatpally, Hyderabad 500 085, India.
e-mail: gklenin@gmail.com

Abstract

In this paper, a novel approach Modified Monkey optimization (MMO) algorithm for solving optimal reactive power dispatch problem has been presented. MMO is a population based stochastic meta-heuristic algorithm and it is inspired by intelligent foraging behaviour of monkeys. This paper improves both local leader and global leader phases. The proposed (MMO) algorithm has been tested in standard IEEE 30 bus test system and simulation results show the worthy performance of the proposed algorithm in reducing the real power loss.

Keywords: *optimal reactive power, transmission loss, modified monkey optimization, bio-inspired algorithm*

1. Introduction

Reactive power optimization plays a key role in optimal operation of power systems. Many numerical methods [1-7] have been applied to solve the optimal reactive power dispatch problem. The problem of voltage stability plays a strategic role in power system planning and operation [8]. So many Evolutionary algorithms have been already proposed to solve the reactive power flow problem [9-11]. In [12, 13], Hybrid differential evolution algorithm and Biogeography Based algorithm has been projected to solve the reactive power dispatch problem. In [14, 15], a fuzzy based technique and improved evolutionary programming has been applied to solve the optimal reactive power dispatch problem. In [16, 17] nonlinear interior point method and pattern based algorithm has been used to solve the reactive power problem. In [18-20], various types of probabilistic algorithms utilized to solve optimal reactive power problem. This paper introduces a novel Modified Monkey optimization for solving optimal reactive power dispatch power problem. Monkey Optimization algorithm [21] is fresh entry in class of swarm intelligence. This Monkey Optimization algorithm is enthused by fission fusion social structure based on foraging behaviour of monkeys when searching for quality food source and for mating. Alike to any other population based optimization techniques, artificial bee colony (ABC) consists of a population of intrinsic solutions. The intrinsic solutions are food sources of honey bees. The fitness is decided in terms of the quality of the food source that is nectar amount. Artificial bee colony is comparatively a direct, quick and population based stochastic exploration technique in the field of nature inspired algorithms. Monkey Optimization algorithm is also alike to ABC in nature. There are two fundamental processes which drive the swarm to modernize in ABC: the deviation process, which empowers exploring different fields of the exploration space, and the selection process, which guarantees the exploitation of the preceding experience. However, it has been shown that the ABC may infrequently stop moving toward the global optimum even though the population has not meeting to a local optimum [22]. It can be observed that the solution search equation of ABC algorithm is good at exploration but poor at exploitation [23]. Therefore, to uphold the good equilibrium between exploration and exploitation behaviour of ABC, it is extremely expected to develop a local exploration method in the basic ABC to strengthen the exploration region. The proposed MMO algorithm has been evaluated in standard IEEE 30 bus test system & the simulation results show that our proposed approach outperforms all reported algorithms in minimization of real power loss.

2. Problem Formulation
2.1. Active Power Loss
The objective of the reactive power dispatch is to minimize the active power loss in the transmission network, which can be described as follows:

$$F = PL = \sum_{k \in Nbr} g_k \left(V_i^2 + V_j^2 - 2V_i V_j \cos\theta_{ij} \right) \tag{1}$$

Or,

$$F = PL = \sum_{i \in Ng} P_{gi} - P_d = P_{gslack} + \sum_{i \neq slack}^{Ng} P_{gi} - P_d \tag{2}$$

Where g_k : is the conductance of branch between nodes i and j, N_{br}: is the total number of transmission lines in power systems. P_d: is the total active power demand, P_{gi}: is the generator active power of unit i, and P_{gsalck}: is the generator active power of slack bus.

2.2. Voltage Profile Improvement
For minimizing the voltage deviation in PQ buses, the objective function becomes:

$$F = PL + \omega_v \times VD \tag{3}$$
Where ωv: is a weighting factor of voltage deviation.

VD is the voltage deviation given by:

$$VD = \sum_{i=1}^{Npq} |V_i - 1| \tag{4}$$

2.3. Equality Constraint
The equality constraint of the optimal reactive power dispatch power (ORPD) problem is represented by the power balance equation, where the total power generation must cover the total power demand and the power losses:

$$P_G = P_D + P_L \tag{5}$$

This equation is solved by running Newton Raphson load flow method, by calculating the active power of slack bus to determine active power loss.

2.4. Inequality Constraints

The inequality constraints reflect the limits on components in the power system as well as the limits created to ensure system security. Upper and lower bounds on the active power of slack bus, and reactive power of generators:

$$P_{gslack}^{min} \leq P_{gslack} \leq P_{gslack}^{max} \tag{6}$$

$$Q_{gi}^{min} \leq Q_{gi} \leq Q_{gi}^{max}, i \in N_g \tag{7}$$

Upper and lower bounds on the bus voltage magnitudes:

$$V_i^{min} \leq V_i \leq V_i^{max}, i \in N \tag{8}$$

Upper and lower bounds on the transformers tap ratios:

$$T_i^{min} \leq T_i \leq T_i^{max}, i \in N_T \tag{9}$$

Upper and lower bounds on the compensators reactive powers:

$$Q_c^{min} \leq Q_c \leq Q_C^{max}, i \in N_c \tag{10}$$

Where N is the total number of buses, NT is the total number of Transformers; Nc is the total number of shunt reactive compensators.

3. Monkey Optimization Algorithm

JC Bansal et al. [21] used Social activities of monkeys to develop a stochastic optimization algorithm that impersonate fission-fusion social structure based intelligent foraging behaviour of spider monkeys. JC Bansal et al. [21] identified four key features.

Step 1. The group starts food foraging and evaluates their distance from the food.

Step 2. Group members update their positions based on the distance from the foods source and all over again evaluate distance from the food sources.

Step 3. Moreover, in this step, the local leader modernizes its best location within the group and if the location is not rationalized for a predefined number of times then all members of that group start searching of the food sources in different directions.

Step 4. Consequently, in the last step, the global leader keep informed its eternally best position and in case of inactivity, it divides the group into smaller size subgroups.

It is witnessed that in SMO algorithm [21] there are two most significant control parameters are Global Leader Limit (GLlimit) and Local Leader Limit (LLlimit) which provide suitable direction to global and local leaders respectively. In SMO immobility can be evaded by using LLlimit. If a local group leader does not keep informed her-self after a predefined number of times then that group is re-directed to another direction for in order to quest for food. Here, the term predefined number of times is referred as LLlimit. An additional control parameter, that is to say Global Leader Limit (GLlimit) is used for the identical intention by global leader. The global leader divides the group into smaller sub-groups if she does not modernize in a predefined number of times that is GLlimit. Similar to the other population-based algorithms, SMO is also a hit and trial based mutual iterative approach. The SMO development consists of seven major phases. The detailed description of each step of SMO achievement is delineated below:

a) Initialization of the Population

At the outset, SMO [21] produces an consistently dispersed primary population of N spider monkeys where each monkey SMi (i = 1, 2, ..., N) is a vector of dimension D. At this point D is the number of variables in the optimization problem and SMi represent the position of ith Spider Monkey (SM) in the population. Each spider monkey SM corresponds to the potential solution of the problem under consideration. Each SMi is initialized as follows:

$$SM_{ij} = SM_{minj} + \emptyset \times (SM_{maxj} - SM_{minj}) \tag{11}$$

b) Local Leader Phase (LLP)

The second phase in SMO is Local Leader phase. In this phase SM update its existing location based on the information from the local leader understanding as well as local group members understanding. The fitness value of so obtained new location is calculated. If the fitness value of the new location is higher than that of the previous location, subsequently the SM updates his location with the new one. The location modernizes equation for ith SM (which is a member of kth local group) in this phase is as follow:

$$SM_{newij} = SM_{ij} + \emptyset_1 \times (LL_{kj} - SM_{ij}) + \emptyset_2 \times (SM_{rj} - SM_{ij}) \tag{12}$$
$$\text{where } \emptyset_1 \in (0,1) \text{and } \emptyset_2 \in (-1,1)$$

Where SMij is the jth dimension of the ith SM, LLkj represents the jth dimension of the kth local group leader position. SMrj is the jth dimension of the rth SM which is chosen randomly within kth group such that r≠ i.

c) Global Leader Phase (GLP)

After achievement of the Local Leader phase next phase is Global Leader phase (GLP). During GLP phase, all the SM's bring up to date their location by means of understanding of Global Leader and local group member's understanding. The location modernizes equation for this phase is as follows:

$$SM_{newij} = SM_{ij} + \emptyset_1 \times \left(GL_j - SM_{ij}\right) + \emptyset_2 \times \left(SM_{rj} - SM_{ij}\right) \qquad (13)$$
where $\emptyset_1 \in (0,1)$ and $\emptyset_2 \in (-1,1)$.

Where GLj stands for the jth dimension of the global leader location and $j \in \{1, 2, ...,D\}$ is the haphazardly preferred index.

In GLP phase, the locations of spider monkeys (SMi) are modernized based on probabilities pi's which are considered using their fitness. In this way a better candidate will have more chance to make itself better. The probability pi may be calculated using following expression:

$$p_i = 0.9 \times \frac{fitness_i}{fitness_{max}} + 0.1 \qquad (14)$$

Here fitnessi is the fitness value of the ith SM and fitnessmax is the maximum fitness in the group. Further, the fitness of the newly produced position of the SM's is calculated and compared with the old one and adopted the better position.

d) Global Leader Learning (GLL) phase

In GLL phase, the location of the global leader is restructured by applying the rapacious selection approach in the population i.e., the location of the SM having most outstanding fitness in the population is selected as the modernized location of the global leader. Further, it is checked that the location of global leader is modernizing or not and if not then the Global Limit Count is incremented by 1.

e) Local Leader Learning (LLL) phase

In LLL phase, the position of the local leader is modernized by applying the greedy selection in that group i.e., the location of the SM having unmatched fitness in that group is preferred as the updated location of the local leader. Next, the updated location of the local leader is compared with the older one and if the local leader is not updated then the Local Limit Count is incremented by 1.

f) Local Leader Decision (LLD) phase

If any Local Leader position is not modernized up to a predefined threshold called Local Leader Limit (LLLimit), then all the members of that group update their locations either by arbitrary initialization or by using mutual information from Global Leader and Local Leader through equation (15), based on the pr (perturbation rate).

$$SM_{newij} = SM_{ij} + \emptyset \times \left(GL_j - SM_{ij}\right) + \emptyset \times \left(SM_{ij} - LL_{kj}\right) \qquad (15)$$
where $\emptyset \in (0,1)$

It is comprehensible from the equation (15) that the modernized measurement of this SM is attracted towards global leader and fends off from the local leader.

g) Global Leader Decision (GLD) phase

In GLD phase, the position of global leader is observed and if it is not updated up to a predefined number of iterations is known as Global Leader Limit (GLlimit), then the global leader divides the population into smaller groups. Firstly, the population is divided into two groups and then three groups and so on till the maximum number of groups (MNG) are formed. Each time in GLD phase, LLL procedure is began to decide on the local leader in the recently fashioned groups. The case in which maximum number of groups is formed and even then the position of global leader is not modernized then the global leader pools all the groups to form a single group. As a consequence the predicted algorithm mimics fusion-fission structure of SMs. The complete pseudo-code of the SMO algorithm is outlined as follow [21]:

Spider Monkey Optimization (SMO) Algorithm:

Step 1. Set Population, Local Leader Limit (LLlimit), Global Leader Limit (GLlimit) and Perturbation rate (pr).

Step 2. Calculate fitness (The distance of each individual from corresponding food sources).

Step 3. Select leaders (global and local both) by smearing greedy selection.

Step 4. while (Annihilation criteria is not fulfilled) do

Step 5. Produce the new locations for all the group members by using self-experience, local leader experience and group member's experience. Using Equation (12)

Step 6. Apply the gluttonous selection process between existing location and newly generated location, based on fitness and select the enhanced one.

Step 7. Calculate the probability pi for all the group members using Equation (14).

Step 8. Produce new locations for the all the group members, selected by pi, by using self-experience, global leader experience and group members experiences Using Equation (13)

Step 9. Modernize the position of local and global leaders, by applying the greedy selection process on all the groups.

Step 10. If any Local group leader is not updating her position after a specified number of times (LLLimit) then re-direct all members of that particular group for foraging by algorithm using Equation (15).

Step 11. If Global Leader is not modernizing her position for a specified number of times (GLLimit) then she divides the group into smaller groups by following steps.

Step 12. End While

4. Modified Monkey Optimization (MMO) Algorithm

Exploration of the whole search space and exploitation of the best solutions found in its proximity may be balanced by maintaining the diversity in local leader and global leader phase of SMO. In order to balance exploration and exploitation of local search space the proposed algorithm alter both local leader phase and global leader phase using modified golden section search (GSS) [24]. Original GSS method does not use any gradient information of the function to finds the optima of a uni-modal continuous function. GSS processes the interval [a = −1.2, b = 1.2] and initiates two intermediate points:

$$F1 = b - (b - a) \times \psi,$$ (16)

$$F2 = a + (b - a) \times \psi,$$ (17)

Here $\psi = 0.618$ is the golden ratio.
The detailed GSS process [25] is described as follows:

Golden Section Search procedure
Input: Optimization function
Min f(x) s.t. a ≤ x ≤ b and termination criteria
Repeat while termination criteria fulfill
Calculate F1and F2 as follow
F1=b-(b-a)*Ψ and F2=a+(b-a)*Ψ here a = −1.2, b = 1.2 and Ψ=0.618(Golden ratio)
Compute f(F1) and f(F2)
If f(F1)< f(F2) then
b = F2 and the solution fall in range [a,b]
else
a = F1 and the solution fall in range [a,b]
end if
end while

The proposed strategy modify Equation (12) and (13) in the following manner. Here f is determined by GSS process as outlined in above algorithm. Position update in local leader phase is done using Equation (18).

$$SM_{newij} = SM_{ij} + \emptyset_1 \times \left(LL_{kj} - SM_{ij}\right) + \emptyset_2 \times \left(SM_{rj} - SM_{ij}\right) + f \times \left(SM_{rj} - SM_{ij}\right)$$ (18)
here $\emptyset_1 \in (0,1)$ and $\emptyset_2 \in (-1,1)$ f is decided by GSS.

Position update in local leader phase is done using Equation (19).

$$SM_{newij} = SM_{ij} + \emptyset_1 \times \left(GL_j - SM_{ij}\right) + \emptyset_2 \times \left(SM_{rj} - SM_{ij}\right) + f \times \left(SM_{rj} - SM_{ij}\right)$$ (19)
here $\emptyset_1 \in (0,1)$ and $\emptyset_2 \in (-1,1)$ f is decided by GSS.

MMO Algorithm:

Step 1. Initialize Population, Local Leader Limit (LLlimit), Global Leader Limit (GLlimit) and Perturbation rate (pr).

Step 2. Compute fitness (The distance of each individual from corresponding food sources).

Step 3. Select leaders (global and local both) by applying greedy selection.

Step 4. while (extermination criteria is not fulfilled) do

Step 5. For finding the objective (Food Source), generate the new locations for all the group members by using self-experience, local leader experience and group member's experience.

$$SM_{newij} = SM_{ij} + \emptyset_1 \times \left(LL_{kj} - SM_{ij}\right) + \emptyset_2 \times \left(SM_{rj} - SM_{ij}\right) + f \times \left(SM_{rj} - SM_{ij}\right)$$

where $\emptyset_1 \in (0,1)$ and $\emptyset_2 \in (-1,1)$ f is decided by GSS.

Step 6. Apply the gluttonous selection process between existing location and newly generated location, based on fitness and select the better one;

Step 7. Calculate the probability pi for all the group members using,

$$p_i = 0.9 \times \frac{fitness_i}{fitness_{max}} + 0.1$$

Step 8. Produce new locations for the all the group members, selected by pi, by using self-experience, global leader experience and group member's experiences.

$$SM_{newij} = SM_{ij} + \emptyset_1 \times \left(GL_j - SM_{ij}\right) + \emptyset_2 \times \left(SM_{rj} - SM_{ij}\right) + f \times \left(SM_{rj} - SM_{ij}\right)$$

where $\emptyset_1 \in (0,1)$ and $\emptyset_2 \in (-1,1)$ f is decided by GSS.

Step 9. Modernize the position of local and global leaders, by applying the greedy selection process on all the groups.

Step 10. If any Local group leader is not updating her position after a specified number of times (LLLimit) then re-direct all members of that particular group for foraging by algorithm.

if $U(0,1) \geq pr$
$$SM_{newij} = SM_{minj} + \emptyset \times \left(SM_{maxj} - SM_{minj}\right)$$
Else
$$SM_{newij} = SM_{ij} + \emptyset \times \left(GL_j - SM_{ij}\right) + \emptyset \times \left(SM_{ij} - LL_{kj}\right)$$
where $\emptyset \in (0,1)$

Step 11. If Global Leader is not updating her position for a specified number of times (GLLimit) then she divides the group into smaller groups by following steps.

if Global Limit Count > GLLimit then set Global Limit Count = 0
if Number of groups < MNG then
Divide the population into groups.
else
Pool all the groups to make a single group.
Modernize Local Leaders position.

5. Simulation Results

MMO algorithm has been tested on the IEEE 30-bus, 41 branch system. It has a total of 13 control variables as follows: 6 generator-bus voltage magnitudes, 4 transformer-tap settings, and 2 bus shunt reactive compensators. Bus 1 is the slack bus, 2, 5, 8, 11 and 13 are taken as PV generator buses and the rest are PQ load buses. The measured security constraints are the voltage magnitudes of all buses, the reactive power limits of the shunt VAR compensators and the transformers tap settings limits. The variables limits are listed in Table 1.

The transformer taps and the reactive power source installation are discrete with the changes step of 0.01. The power limits generators buses are represented in Table 2. Generators buses are: PV buses 2, 5, 8, 11, 13 and slack bus is 1. The others are PQ-buses.

Table 1. Initial Variables Limits (PU)

Control variables	Min.value	Max.value	Type
Generator: Vg	0.92	1.08	Continuous
Load Bus: VL	0.90	1.01	Continuous
T	0.90	1.40	Discrete
Qc	-0.11	0.30	Discrete

Table 2. Generators Power Limits in MW and MVAR

Bus n°	Pg	Pgmin	Pgmax	Qgmin
1	97.00	50	200	-20
2	80.00	20	80	-20
5	52.00	15	55	-13
8	20.00	10	31	-13
11	20.00	10	25	-10
13	20.00	11	40	-13

Table 3 show that the proposed approach succeeds in keeping the dependent variables within their limits. Table 4 summarizes the results of the optimal solution by different methods. It reveals the reduction of real power loss after optimization.

Table 3. Values of Control Variables after Optimization and Active Power Loss

Control Variables (p.u)	MMO
V1	1.0308
V2	1.0379
V5	1.0190
V8	1.0289
V11	1.0619
V13	1.0428
T4,12	0.00
T6,9	0.01
T6,10	0.91
T28,27	0.90
Q10	0.11
Q24	0.10
PLOSS	4.5328
VD	0.9081

Table 4. Comparison Results of Different Methods

Methods	Ploss (MW)
SGA (26)	4.98
PSO (27)	4.9262
LP (28)	5.988
EP (28)	4.963
CGA (28)	4.980
AGA (28)	4.926
CLPSO (28)	4.7208
HSA (29)	4.7624
BB-BC (30)	4.690
MMO	4.5328

6. Conclusion

In this paper, the MMO has been successfully implemented to solve Optimal Reactive Power Dispatch problem. The main advantages of the MMO are easily handling of non-linear constraints. The proposed algorithm has been tested on the IEEE 30-bus system to minimize the active power loss. The optimal setting of control variables are well within the limits. The results were compared with the other heuristic methods and proposed MMO demonstrated its effectiveness and robustness in minimizing the real power loss.

References

[1] O Alsac, B Scott. Optimal load flow with steady state security. *IEEE Transaction*. 1973; 745-751.
[2] Lee KY, Paru YM, Oritz JL. A united approach to optimal real and reactive power dispatch. *IEEE Transactions on power Apparatus and systems*. 1985; 104: 1147-1153.
[3] A Monticelli, MVF Pereira, S Granville. Security constrained optimal power flow with post contingency corrective rescheduling. *IEEE Transactions on Power Systems*. 1987; 2(1): 175-182.
[4] Deeb N, Shahidehpur SM. Linear reactive power optimization in a large power network using the decomposition approach. *IEEE Transactions on power system*. 1990; 5(2): 428-435.
[5] E Hobson. Network consrained reactive power control using linear programming. *IEEE Transactions on power systems*. 1980; 99(4): 868-877.
[6] KY Lee, YM Park, JL Oritz. *Fuel –cost optimization for both real and reactive power dispatches*. IEE Proc. 131C(3): 85-93.

[7]   MK Mangoli. KY Lee. Optimal real and reactive power control using linear programming. *Electr.Power Syst.Res.* 1993; 26: 1-10.

[8]   CA Canizares, ACZ de Souza, VH Quintana. Comparison of performance indices for detection of proximity to voltage collapse. 1996; 11(3): 1441-1450.

[9]   SR Paranjothi, K Anburaja. Optimal power flow using refined genetic algorithm. *Electr.Power Compon.Syst.* 2002; 30: 1055-1063.

[10] D Devaraj, B Yeganarayana. *Genetic algorithm based optimal power flow for security enhancement.* IEE proc-Generation Transmission and Distribution. 2005; 152.

[11] A Berizzi, C Bovo, M Merlo, M Delfanti. A ga approach to compare orpf objective functions including secondary voltage regulation. *Electric Power Systems Research.* 2012; 84(1) 187–194.

[12] CF Yang, GG Lai, CH Lee, CT Su, GW Chang. Optimal setting of reactive compensation devices with an improved voltage stability index for voltage stability enhancement. *International Journal of Electrical Power and Energy Systems.* 2012; 37(1): 50–57.

[13] P Roy, S Ghoshal, S Thakur. Optimal var control for improvements in voltage profiles and for real power loss minimization using biogeography based optimization. *International Journal of Electrical Power and Energy Systems.* 2012; 43(1): 830–838.

[14] B Venkatesh, G Sadasivam, M Khan. A new optimal reactive power scheduling method for loss minimization and voltage stability margin maximization using successive multi-objective fuzzy lp technique. *IEEE Transactions on Power Systems.* 2000; 152: 844–851.

[15] W Yan, S Lu, D Yu. A novel optimal reactive power dispatch method based on an improved hybrid evolutionary programming technique. *IEEE Transactions on Power Systems.* 2004; 19(2): 913–918.

[16] W Yan, F Liu, C Chung, K Wong. A hybrid genetic algorithminterior point method for optimal reactive power flow. *IEEE Transactions on Power Systems.* 2006; 21(3): 1163–1169.

[17] J Yu, W Yan, W Li, C Chung, K Wong. An unfixed piecewiseoptimal reactive power-flow model and its algorithm for ac-dc systems. *IEEE Transactions on Power Systems.* 2008; 23(1): 170–176.

[18] F Capitanescu. Assessing reactive power reserves with respect to operating constraints and voltage stability. *IEEE Transactions on Power Systems.* 2011; 26(4): 2224–2234.

[19] Z Hu, X Wang, G Taylor. Stochastic optimal reactive power dispatch: Formulation and solution method. *International Journal of Electrical Power and Energy Systems.* 2010; 32(6): 615–621.

[20] A Kargarian, M Raoofat, M Mohammadi. Probabilistic reactive power procurement in hybrid electricity markets with uncertain loads. *Electric Power Systems Research.* 2012; 82(1): 68–80.

[21] JC Bansal, H Sharma, SS Jadon, M Clerc. Spider Monkey Optimization algorithm for numerical optimization. *Memetic Computing.* 2013; 1-17.

[22] D Karaboga et al. A comparative study of artificial bee colony algorithm. *Applied Mathematics and Computation.* 2009; 214(1):108–132.

[23] G Zhu et al. Gbest-guided artificial bee colony algorithm for numerical function optimization. *Applied Mathematics and Computation.* 2010; 217(7):3166–3173.

[24] J Kiefer. *Sequential minimax search for a maximum.* Proceedings of American Mathematical Society. 1953; 4: 502–506.

[25] JC Bansal, H Sharma, KV Arya, A Nagar. Memetic search in artificial bee colony algorithm. *Soft Computing.* 2013; 1-18.

[26] QH Wu, YJ Cao, JY Wen. Optimal reactive power dispatch using an adaptive genetic algorithm. *Int. J. Elect. Power Energy Syst.* 1998; 20: 563-569.

[27] B Zhao, CX Guo, YJ CAO. Multiagent-based particle swarm optimization approach for optimal reactive power dispatch. *IEEE Trans. Power Syst.* 2005; 20(2): 1070-1078.

[28] Mahadevan K, Kannan PS. Comprehensive Learning Particle Swarm Optimization for Reactive Power Dispatch. *Applied Soft Computing.* 2010; 10(2): 641–52.

[29] AH Khazali, M Kalantar. Optimal Reactive Power Dispatch based on Harmony Search Algorithm. *Electrical Power and Energy Systems.* 2011; 33(3) 684–692.

[30] S Sakthivel, M Gayathri, V Manimozhi. A Nature Inspired Optimization Algorithm for Reactive Power Control in a Power System. *International Journal of Recent Technology and Engineering (IJRTE).* 2013; 2(1): 29-33.

6

Design of Solar PV Cell based Inverter for Unbalanced and Distorted Industrial Loads

D. V. N Ananth[1], G. V. Nagesh Kumar[2]
[1]Viswanadha Institute of Technology and Management, Visakhapatnam
[2]GITAM University, Visakhapatnam
e-mail: drgvnk14@gmail.com

Abstract

PV cell is getting importance in low and medium power generation due to easy installation, low maintenance and subsidies in price from respective nation. Most of the loads in distribution system are unbalanced and distorted, due to which there will be unbalanced voltage and current occur at load and may disturb its overall performance. Due to these loads voltage unbalance, distorted voltage and current and variable power factors in each phase can be observed. An efficient algorithm to mitigate unbalanced and distorted load and source voltage and current in solar photo voltaic (PV) inverter for isolated load system was considered. This solar PV system can be applicable to remote located industrial loads like heating, welding and small arc furnace type distorted loads and also for unbalanced loads. The PV inverter is designed such that it will maintain nearly constant voltage magnitude and can mitigate harmonics in voltage and current near the load terminals. A MATLAB/ SIMULINK based solar PV inverter was simulated and results are compared with standard AC three phase grid connected system. The proposed shows that the inverter is having very less voltage and current harmonic content and can maintain nearly constant voltage profile for highly unbalanced system.

Keywords: distorted load, unbalanced load, PWM inverter, total harmonic distortion (THD), voltage mitigation

1. Introduction

Photovoltaic power generation is getting more significance over past decade. This increase can be expected due to the major factor like greenhouse effect, rapidly increasing fossil fuel price and diminishing resources. In tropical countries like India, sunlight is available for more than half of the time of the day; installation of solar cells is highly recommendable. It was observed that roof top solar panels are installed for water heating, reliable and economic power supply was increased in last five years [1-5]. The main advantages of solar PV cells are, it can be placed in the corner of the building where much sunlight is available; it requires very less maintenance and have more running life, excess energy can be stored in battery or can be pumped to grid.

The solar PV with battery energy storage system is helpful for maintaining continuous and reliable power supply to isolated agriculture type loads [6]. The solar panels can be applicable to residential loads by embedding on the roof top [7-9]. The solar panels are used for applications like road transport lighting and in electric power distribution network [10, 11]. There has been very less work been proposed for application for industrial loads like arc furnace etc. Application of PV cell for non-linear loads and its harmonics analysis was analyzed in [12, 13]. A three-phase DC to AC inverters with high efficiency, low cost, enhanced reliability are designed in [14- 17].

The present paper extended the application of photovoltaic cell to non-linear and unbalanced industrial AC loads by using DC to AC inverter. The PWM based inverter is designed to maintain nearly constant voltage profile for unbalanced load and harmonic mitigation for distorted non-linear diode rectifier RL load. In this, voltage and current unbalance and distortion values to be within limits and with optimum power factor value with inductance or capacitance loads. Equations for unbalanced load and source voltage and current with average power to be delivered by grid source or PV inverter are derived.

The paper describes modeling of the PV inverter system for distorted and unbalanced loads, mitigation of voltage and current harmonics with analytical explanation and control circuit of inverter was explained in forth coming sections. Later, simulation results are given and conclusion with appendix and references were presented.

2. Solar PV Based Inverter for Distorted and Unbalanced Loads

In general, small scale industrial loads in developing countries like India are welding, induction heating and small rating arc furnace loads. Most of the loads in the same industry or industrial locality are highly unbalanced. If these industries are located in remote places or rural areas, they may not get quality power supply and also power lasts for few hours during summer due to many reasons. In order to have reliable power, if industry generates their power by installing a captive power plant with solar photovoltaic (PV) system is a better alternative.

The solar photovoltaic based inverter for distorted and unbalanced isolated industrial loads is shown in Figure 1. The PV cell is a current source device with Dc supply as output. In order to extract maximum power from solar cell, maximum power point tracking (MPPT) algorithm is used. The 19 unique techniques for MPPT algorithm can be available [20]. The DC output voltage from PV cell is to be converted to AC as most of the loads are of AC supply. Capacitor near the bidirectional inverter is used to maintain nearly constant voltage and also acts as a reactive power supplier for maintain nearly constant voltage profile.

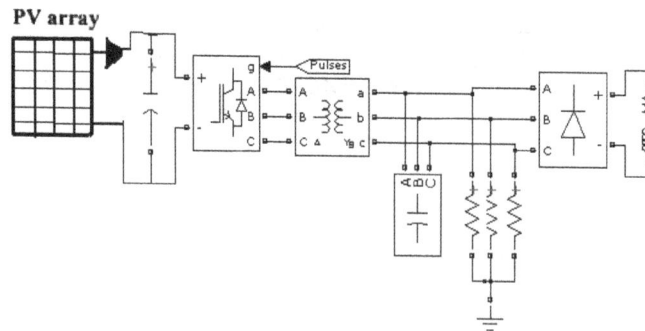

Figure 1. PV cell based inverter for distorted and unbalanced loads

The transformer is used for step-up or step-down voltage to meet voltage demand by the load. The star-delta combination of transformer is also used for suppressing third order (3n) harmonics and star point neutral grounding for protection, measurement and grounding unbalanced current. The diode rectifier with resistive (R) or resistive- inductive (RL) loads represents distorted loads like arc furnace, induction heating (low resistance, high inductance coil) or as welding type AC loads. The shunt type resistive loads if they are unequal, they represent unbalanced loads. The unbalanced load voltages can be represented as:

$$V_{lA}=V_{lmA}\sin(\theta) \tag{1}$$

$$V_{lB}=V_{lmB}in(\theta-120^0+b) \tag{2}$$

$$V_{lC}=V_{lmC}\sin(\theta+120^0+c) \tag{3}$$

Here V_{lA}, V_{lB} and V_{lC} are load voltages in A, B and C phases with RMS voltages in each phases A, B and C are V_{lmA}, V_{lmB} and V_{lmC}.

In this each phase difference is 1200 between them and due to unbalanced system; there is small voltage angle shift in phases B and C with small angles b and c. if the system is perfectly balanced, angles b and will be zero. The load currents are given by below equations:

$$I_{lA}=I_{lmA}\sin(\theta-\Phi) \tag{4}$$

$$I_{lB}=I_{lmB}\sin(\theta-120^0+b-\Phi) \tag{5}$$

$$I_{lC} = I_{lmC}\sin(\theta + 120^0 + c - \Phi) \qquad (6)$$

Here I_{lA}, I_{lB} and I_{lC} are load voltages in A, B and C phases with RMS voltages in each phases A, B and C are I_{lmA}, I_{lmB} and I_{lmC}.

The angle Φ is power factor angle which will exist if the load contains resistive with inductance and or capacitance value. For unity power factor load with only resistance the angle Φ will be zero. The average power delivered by three phase source or inverter can be written as:

$$P_{l_{avg}} = \frac{1}{2}[V_{lA}I_{lA} + V_{lB}I_{lB} + V_{lC}I_{lC}]\cos(\Phi) \qquad (7)$$

Substituting equations 1, 2, 3, 4, 5 and 6 in equation 7 we get average power to be supplied by three phase source or inverter supply.

The inverter or three phase supply current is given three equations as:

$$I_{sA} = \frac{2P_{l_{avg}}V_{lA} \llcorner -\Phi}{\cos\Phi V_{lmA}(V_{lmA} + V_{lmB} + V_{lmC})} \qquad (8)$$

$$I_{sB} = \frac{2P_{l_{avg}}V_{lb} \llcorner -\Phi}{\cos\Phi V_{lmB}(V_{lmA} + V_{lmB} + V_{lmC})} \qquad (9)$$

$$I_{sC} = \frac{2P_{l_{avg}}V_{lC} \llcorner -\Phi}{\cos\Phi V_{lmC}(V_{lmA} + V_{lmB} + V_{lmC})} \qquad (10)$$

The total three phase voltage to be supplied by PV cell based inverter is given by:

$$V_i = \frac{1}{2}[V_{iA}\alpha_A + V_{iB}\alpha_B + V_{iC}\alpha_C] \qquad (11)$$

The three phase inverter voltages due to the existence of unbalanced and distorted loads are V_{iA}, V_{iB} and V_{iC} and the compensation voltage coefficients due to unbalanced and distorted loads are α_a, α_b and α_c.

$$\alpha_a = 1, \alpha_b = \frac{\cos(\Phi + b)}{\cos(\Phi)}, \alpha_c = \frac{\cos(\Phi + c)}{\cos(\Phi)} \qquad (12)$$

After perfect compensation b and c angles will become zero and α_b and α_c to be unity can be achieved by PV inverter topology. The load current with all the harmonics included can be represented as:

$$I_l = \sqrt{\sum_{n=1}^{\infty}[(I_{lA}^{2n\pm1})^2 + (I_{lB}^{2n\pm1})^2 + (I_{lc}^{2n\pm1})^2]} \qquad (13)$$

Where n is the order of harmonics, with n=1 as fundamental. In this 3n and 2n order harmonics are cancelled due to symmetrical three phase supply. 2n±1 harmonics are present and based on control circuit and passive filters combination, higher order harmonics can be eliminated. The sum of all harmonics divided by fundamental harmonic gives total harmonic distortion (THD).

3. Mitigation of Voltage and Current Harmonics for Distorted and Unbalanced Industrial Loads

The rectified RL load will distort three phase voltage and current waveforms and linear R load of unbalanced load results in unbalanced voltage and current for Figure 1 can be written as:

$$L\frac{di_c}{dt} = -v_c + \frac{V_{pv}}{2} - L\frac{di_{load}}{dt} \qquad (14)$$

$$L\frac{dv_c}{dt} = -i_c \qquad (15)$$

$$i_m = \mathcal{P}i_c + \mathcal{Q}i_{load} \qquad (16)$$

$$L\frac{di_m}{dt} = -\mathcal{P}v_c + \frac{\mathcal{P}v_{pv}}{2} - (\mathcal{P}-\mathcal{Q})L\frac{di_{load}}{dt} \tag{17}$$

$$\mathcal{P}C\frac{dv_c}{dt} = i_m - \mathcal{Q}i_{load} \tag{18}$$

In the above equations, L represent equivalent inductance of the transformer and smoothing reactor and C is the capacitance of shunt passive filter. V_{pv} is the PV cell output voltage and V_c is shunt filter capacitor voltage. i_m is main current flowing to filter and to the linear and non-linear loads, i_{load} is the current flowing through the two loads. $\mathcal{P}$ and $\mathcal{Q}$ are weights of capacitor current and load currents.

From Equation (1), (3), it can be observed that if load current in all phases are different then capacitance current flowing to filter in respective phases will be different and finally PV cell output voltage and capacitor voltage will be different. The weights ($\mathcal{P}$ and $\mathcal{Q}$) are so chosen to follow Equation (2) and (3), so that (5) is satisfied.

The voltage source inverter is designed such that based on i_m and i_{load} current in each phase, voltage in each leg of the inverter is switched ON or OFF to maintain constant load voltage. Current harmonics are also controlled by using above equations. The voltage harmonics are controlled by using reactive power compensation provided by the inverter side DC capacitor link and proper control strategy to eliminate higher order harmonics and 9th order harmonics are eliminated by passive shunt capacitance filter. Hence among lower order harmonics 5th and 11th are left to be eliminated by using above control strategy.

4. Control Circuit of Inverter

The control circuit for solar PV based inverter for distorted and unbalanced loads is shown in Figure 2.

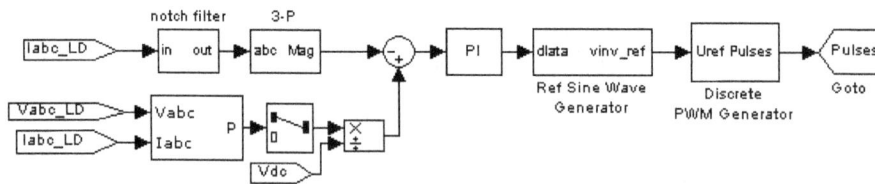

Figure 2. Control circuit for solar PV based inverter for distorted and unbalanced loads

The three phase load current is taken as input and is filtered by using notch filter in order to pass few harmonic order waveforms. The phase currents are converted into magnitude to get actual magnitude of load current. The load real power is divided with PV cell or inverter capacitor voltage to get reference load current. The difference between reference and actual load currents is error in current waveform. This error can be minimized by using optimal tuned proportional and integral controller (PI), which helps also in maintaining stability during rapid switching of pulses and due to other small disturbances and its output is fed to the reference sine wave generator to produce three phase voltage waveform. The circuit for reference sine wave generator is shown in Figure 3.

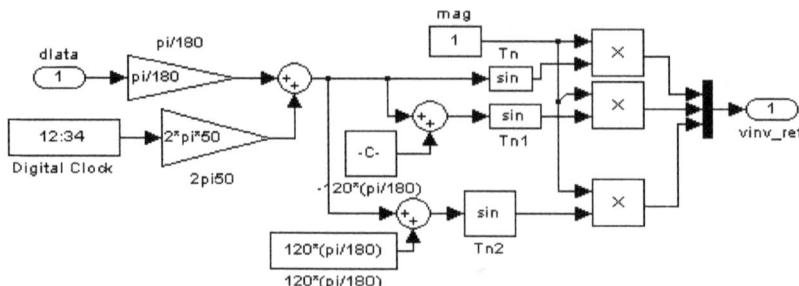

Figure 3. Reference sine wave generator from PI controller output

The three phase reference sine wave is converted into magnitude and angle and is further given to PWM based inverter to generate pulses for voltage source inverter (VSI) which is a bidirectional IGBT switch.

5. Simulation Results

The implemantation of MATLAB/ SIMULINK for grid connected and PV inverter for distorted and unbalanced load is shown in Fig.4 and 5. Now comparison is made between these two circuits for same rated load in case-A and case-B, case-A is grid connected and case-B is with inverter type load. The ratings of the parameters are given in Appendix.

Case-A: Conventional Three phase grid source

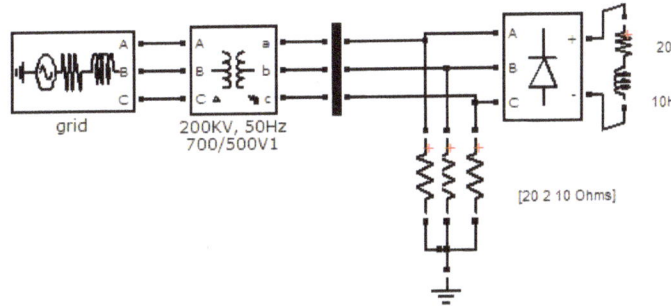

Figure 4. Simulink diagram of grid circuit for distorted and unbalanced load

The load voltage and current waveforms for grid connected distorted and unbalanced loads are shown in Figure 5. It can be observed that all the three phases' voltages are not same, the red color waveform is having 650V, blue is having 300 volts and green is having 550 volts. Similarly red color current waveform is 50A, blue color current wave is 165A and green color phase current is 65A. Both voltage and current are unbalanced.

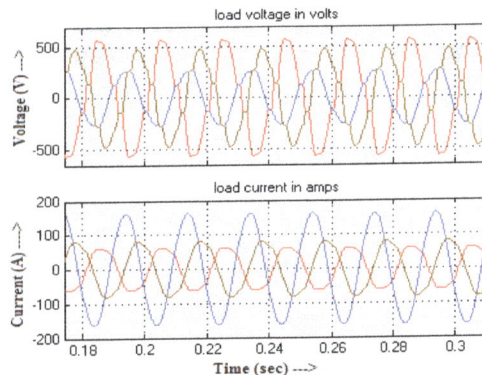

Figure 5. Load voltage and current waveforms for grid supply circuit

The total harmonic distortion (THD) of voltage and current waveform for blue color phase for grid connected non-linear and linear loads are shown in Figure 6 and 7.

It can be observed that voltage THD is 7.25% and current THD is 1.17% for blue phase. For red color phase voltage and current THDs are 4.80 and 6.77%, green color phase is having 6.34 and 3.43%. However it can be observed that power factor is almost unity as resistive load is more dominating than inductive load.

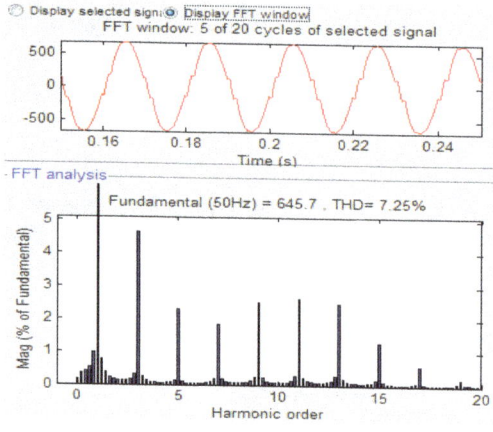

Figure 6. THD of blue color waveform for grid connected load voltage

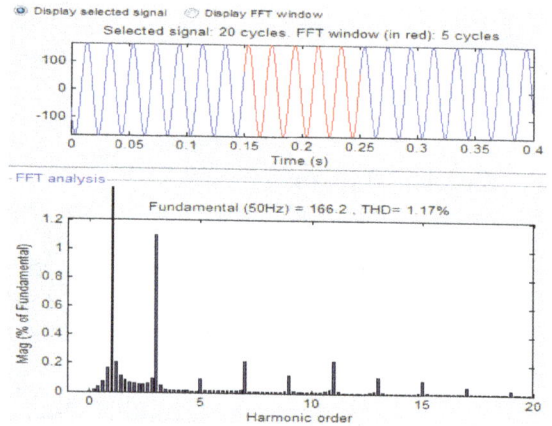

Figure 7. THD of blue color waveform for grid connected load current

Case-B: Inverter Type Load
The PV cell based inverter for same linear and non-linear loads is given in Figure 8.

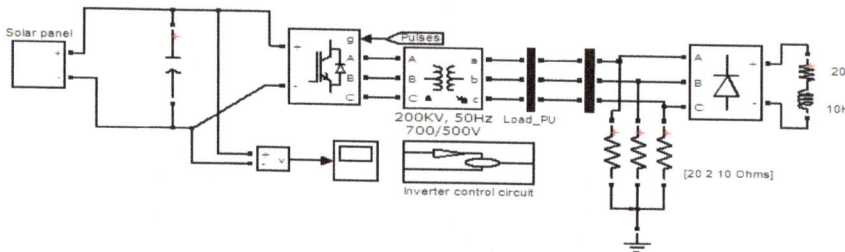

Figure 8. Simulink diagram of PV inverter for distorted and unbalanced load

It can be observed that all the three phase voltages are nearly same and have 650V as peak-to-peak value and current in red phase is 52A, blue phase is 290A and green phase is 82A respectively as shown in Figure 9. The three phase's voltages are 650, 625 and 550 Volts respectively.

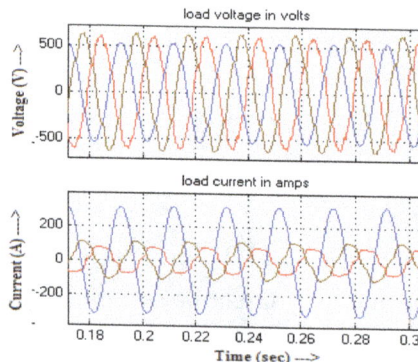

Figure 9. Load voltage and current waveforms for PV inverter circuit

The difference in three phase currents for grid connected and inverter fed loads are different simply due to the fact of voltage unbalance and balance in these cases.

The output voltage, current and power waveforms for solar PV system is shown in Figure 10. In this analysis, the sun irradiation is constant solar cell is working in normal conditions with good cooling and appropriate MPPT algorithm. The output of solar cell is maintained at 900 volts and current varying between 100 and 150Amps. The output power is about 135KW.

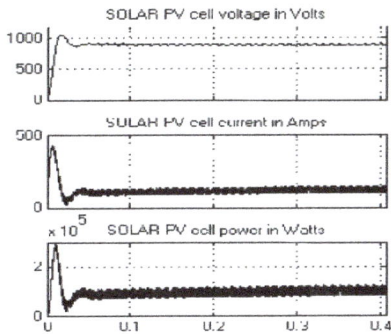

Figure 10. Solar PV cell output voltage, current and power waveforms

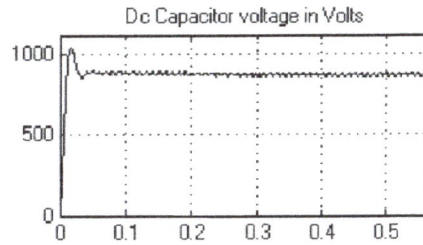

Figure 11. DC link capacitance voltage in volts

The output voltage waveform of DC link capacitor is given in Figure 11 and its value is nearly constant at 850V and the ripples are also minimum. The THD of voltage and current waveform for blue phase is shown in Figure 12 and 13. The load voltage THD is 2.06% and current THD is 1.80% for blue color waveform. The red color wave is having 3.36% and 2.80%, while green color wave is having 4.8% and 1.25% THD. Table 1 shows the output voltage and current with THD values for grid and PV Inverter.

Table 1. Output Voltage and Current With THD Values For Grid And PV Inverter

Ph	3 Phase Grid Supply				PV Cell Inverter			
	V_mag Volts	I_Mag Amps	V_THD %	I_THD %	V_Mag Volts	I_MAG Amps	V_THD %	I_THD %
A	650	50	7.25	6.77	690	52	2.06	2.8
B	550	65	4.80	1.77	625	82	3.36	1.80
C	300	165	6.34	3.43	550	290	4.8	1.25

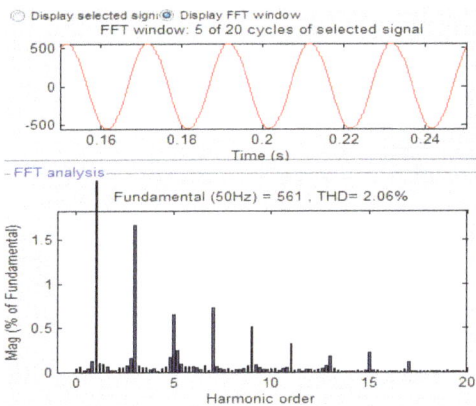

Figure 12. THD of blue color waveform for inverter connected load voltage

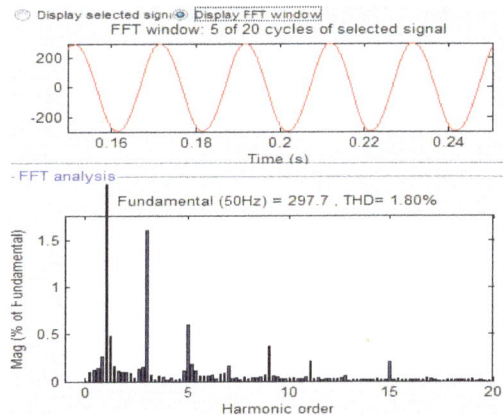

Figure 13. THD of blue color waveform for inverter connected load current

6. Conclusion

For normal grid connected supply, the three phase voltages are different due to unbalanced loads and their voltages are 650, 550 and 300 volts respectively and their voltage THDs are 4.8, 6.34 and 7.16% and current THDs are 6.77, 4.43 and 1.17%. The source or load voltages are unbalanced and have THD in two phases more than 5%, which may not be acceptable. So an additional passive filter is required to mitigate harmonics and active filter to compensate decreased voltage due to load. With the proposed control strategy, the voltage in the three phases is 650, 625 and 550 volts and current is 82, 290 and 52Amps respectively. The same phase's voltage THDs are 2.06, 3.36 and 4.8% while current THDs are 1.8, 2.8 and 1.25%. Hence in all the three phases, maintained nearly constant voltage profile and is also having THD less than 3%. But compared to grid source, current THD is high and in two phases it is more than 3% and however is acceptable to keep a passive filter to decrease this content. The proposed scheme will have 5, 7 and 11 order harmonics in voltage and current waveforms and has to be minimized by using a band pass filter. Therefore our proposed scheme can suppress voltage harmonics due to load and also do not produce much destructive harmonics and is highly capable to maintain load voltage almost constant. DC voltage ripples are minimum and control scheme is easy to implement compared to Park's transformation or phase sequential methodologies.

References

[1] JL Duarte, JAA Wijntjens, J Rozenboom, *Designing light sources for solar-powered systems*. In Proc, 5th European Conf. Power Electronics and Application. 1993; 8: 78–82.
[2] HJ Beukes, JHR Enslin. *Analysis of a new compound converter as MPPT, battery regulator and bus regulator for satellite power systems*. In Proc. IEEE PESC'93. 1993: 846–852.
[3] TF Wu, CH Chang, ZR Liu, TH Yu. *Single-stage converters for photovoltaic powered lighting systems withMPPT and charging features*. In Proc. IEEE APEC'98. 1998: 1149–1155.
[4] DB Snyman, JHR Enslin. *Combined low-cost, high-efficient inverter, peak power tracker and regulator for PV application*. In Proc. IEEE PESC'89. 1989: 67–74.
[5] U Hermann, HG Langer. *Low cost DC to AC converter for photovoltaic power conersion in residential applications*. In Proc. IEEE PESC'93. 1993: 588–594.
[6] DVN Ananth. *Performance evaluation of solar photovoltaic system using maximum power tracking algorithm with battery backup*. In Proc. IEEE T & D'12. 2012: 1–8.
[7] Alam MJE, Muttaqi KM, Sutanto D. Mitigation of Rooftop Solar PV Impacts and Evening Peak Support by Managing Available Capacity of Distributed Energy Storage Systems. *IEEE Trans. Power Syst.* 2013; 28(4): 3874–3884.
[8] Joshi KA, Pindoriya NM. *Impact investigation of rooftop Solar PV system: A case study in India*. In Proc. IEEE ISGT'12. 1993: 1–8.
[9] Alam MJE, Muttaqi KM, Sutanto D. A SAX-Based Advanced Computational Tool for Assessment of Clustered Rooftop Solar PV Impacts on LV and MV Networks in Smart Grid. *IEEE Trans. Smart Grid.* 2013; 4(1): 577–585.
[10] A Canova, L Giaccone, F Spertino, M Tartaglia. Electrical impact of photovoltaic plant in distributed network. *IEEE Trans. Ind.Appl.* 2009; 45(1): 341–347.
[11] M Thomson, DG Infield. Network power-flow analysis for a high penetration of distributed generation. *IEEE Trans. Power Syst.* 2007; 22(3): 1157–1162.
[12] GL Campen. An analysis of the harmonics and power factor effects at a utility intertied photovoltaic system. *IEEE Trans. Power App. Syst.* 1982; 101: 4632–4639.
[13] P Maussion, M Grandpierre, J Faucher, *On the way to real time fuzzy control of a PWM source inverter with nonlinear loads*. In Proc. 5th European Conf. Power Electronics and Application. 1993: 66–71.
[14] B Bolsens, K De Brabandere, J Van den Keybus, J Driesen, R Belmans. *Three-phase observer-based low distortion grid current controller using an LCL output filter*. In Proc. IEEE Conf. Power Electron. Spec. Conf. 2005: 1705–1711.
[15] DB Snyman, JHR Enslin. *Combined low-cost, high-efficient inverter, peak power tracker and regulator for PV application*. In Proc.IEEE PESC'89. 1989: 67–74.
[16] U Hermann, HG Langer. *Low cost DC to AC converter for photovoltaic power conersion in residential applications*. In Proc. IEEE PESC'93. 1993: 588–594.
[17] R Tirumala, P Imbertson, N Mohan, C Henze, R Bonn. *An efficient, low cost DC-AC inverter for photovoltaic systems with increased reliability*. In Proc. IEEE 28th Annu. Conf. Ind. Electron. Soc. 2002; 2: 1095–1100.
[18] RH Bonn. *Developing a 'next generation' PV inverter*. In Proc. 29th IEEE Photovolt. Spec. Conf. 2002: 1352–1355.
[19] X Yuan, Y Zhang. *Status and opportunities of photovoltaic inverters in grid-tied and micro-grid systems*. In Proc. Int. Power Electron. Motion Control Conf. 2006; 1: 1–4.
[20] Trishan Esram, Patrick L Chapman. Comparison of Photovoltaic Array Maximum Power Point Tracking Techniques. *IEEE Trans. Ener.Conv.* 2007; 22(2): 439–449.

7

Development of Microcontroller-Based Ball and Beam Trainer Kit

Gunawan Dewantoro*[1], Deddy Susilo[2], Ditya Clarisa Amanda[3]
Department of Electronics and Computer Engineering, Satya Wacana Christian University,
52-60 Diponegoro Street, Salatiga, Indonesia 50711
e-mail: gunawan.dewantoro@staff.uksw.edu[1], deddy.susilo@ymail.com[2], ditya.clarisa@gmail.com[3]

Abstract

A ball and beam trainer kit based on microcontroller was developed for teaching control system course for the sophomore students. This specially-purposed kit consists of a ball located on a beam with a fixed axle at one of its end. At the other end, a servomotor was employed to control the position of the ball by adjusting the rotation angle of the servomotor. Seven predetermined positions were set to 10, 20, 30, 40, 50, 60, and 70 cm relative to the fixed axle of the beam. The Proportional-Integral-Derivative (PID) scheme was then used to compensate the error. This kit is equipped with a user interface to configure controller coefficients, select the set points, plot the actual ball position, and display parameter values. The user interface program runs on PC or notebook connected to microcontroller via serial communications. A questionnaire-based assessment about the use of this kit was conducted by 17 students taking the course, giving a rating value of 94.12%.

Keywords: ball and beam, control system, PID, user interface

1. Introduction

Traditionally, control systems courses are offered by Electrical and Mechanical Engineering Departments all over the world. As in many other university-level technical courses, interactive tools have been very helpful for the students in order to better comprehend the substances along with formal lectures and exercises. Mostly, an interactive tool in control education provides graphical windows which employ active, dynamic, and clickable components. An interactive tool using Easy Java Simulations was implemented in the field of Control Engineering education [1]. A simulator was developed using user friendly virtual interface software to control the speed of a small size DC motor. The user is able to select and adjust the parameters of any desired controller that is defined and represented virtually [2]. Educative tools, Linear System Analysis and Design (LSAD) [3] and Interactive Learning Modules for Proportional-Integral-Derivative (ILM-PID) [4], were built to make students more active and involved in their own control engineering learning process. Excel spreadsheets for basic control education was developed to help students understand how feedback works [5]. The Ch-Control System Toolkit, a software package for the design and analysis of control system, was utilized to develop Web-based Control System Design and Analysis System (WSCDAS) and Web-based Controller/Compensator Design Module (WCCDM). These software tools were used for teaching automatic control of linear time-invariant system [6]. A control simulation for nonlinear system was applied using JAVA-enabled open source software, therefore, students have all necessary elements to practice using nonlinear system which in this case an inverted pendulum [7].

However, all of the above mentioned works were merely implemented by the use computer-based virtual tools. An actual system needs to be constructed in companion with such virtual counterpart. Microcontroller is a compact microcomputer containing a processor core, memory, input/output peripheral, timer/counter, which is able to govern the operation of any plants. Microcontrollers have also been used in learning control system with laboratory scale practical models. Low-cost digital temperature control kit was designed to implement various control strategies using microcontroller [8]. A microcontroller–based control application was designed to distribute supervisory and control tasks within a plant in industry [9]. Low-cost educational microcontroller-based tool for fuzzy logic-controlled line-following mobile robot was put into practice for teaching second-year students [10].

Ball and Beam is a widespread example of application of feedback control system and has been theoretically taught in most universities. In this paper, an ATMega 8535 microcontroller-based ball and beam simulator was built to encompass both virtual and actual tools for learning control system course. This kit consists of a long beam which can be tilted to control the displacement of a ball on it using Proportional-Integral-Derivative (PID) controller. Hardware details and the user interface implementing the controller are given in the paper.

2. System Modeling

A set of mathematical equations was addressed to represent the ball and beam employing appropriate physical laws. All variables contained in the ball and beam system are shown in Figure 1.

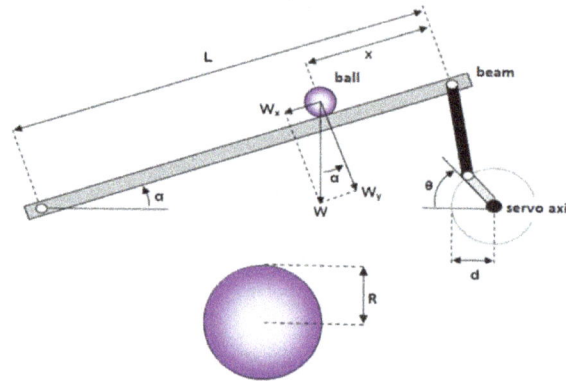

Figure 1. Ball and Beam physical model

Translational acceleration of the ball is given by:

$$\frac{d^2x}{dt^2} = \ddot{x} \tag{1}$$

The exerted force due to translational motion, F_{tx}, is:

$$F_{tx} = m\ddot{x} \tag{2}$$

While the torque of the ball due to rotational motion, T_r, is:

$$T_r = F_{rx}R = J\frac{d\omega_b}{dt} = J\frac{d\left(\frac{V_b}{R}\right)}{dt} = J\frac{d^2\left(\frac{x}{R}\right)}{dt^2} = \frac{J}{R}\ddot{x} \tag{3}$$

where,
F_{rx} = exerted force due to ball rotation
J = moment of inertia of solid ball
ω_b = angular velocity of the ball
v_b = translational velocity of the ball

From (1), (2), and (3) we obtain:

$$F_{rx} = \frac{T_r}{R} = \frac{J}{R^2}\ddot{x} = \frac{\frac{2}{5}mR^2}{R^2}\ddot{x} = \frac{2}{5}m\ddot{x} \tag{4}$$

It is obviously known that W_x is the force due to mass of the ball (m) and gravitational acceleration (g) along the x-axis while neglecting frictions between the ball and the beam, and can be formulated by:

$$\frac{2}{5}m\ddot{x} + m\ddot{x} = mg \sin \alpha \qquad (5)$$

By dividing both sides by m, we have:

$$\frac{5}{7}g \sin \alpha = \ddot{x} \qquad (6)$$

For α small, sin α is approximated by α, giving:

$$\frac{5}{7}g\alpha = \ddot{x} \qquad (7)$$

Meanwhile, the relationship between the angle servomotor axle and the beam is given by:

$$\alpha = \frac{d}{L}\theta \qquad (8)$$

By substituting Equation (8) to (7), we get:

$$\frac{5}{7}g\frac{d}{L}\theta = \ddot{x} \qquad (9)$$

Therefore, the transfer function of the ball and beam in s-domain is given by:

$$\frac{X(s)}{\theta(s)} = \frac{5gd}{7L}\frac{1}{s^2} \qquad (10)$$

Constants g, d, and L are 9.8 m/s², 5 cm , and 69 cm, consecutively. It is very clear from the above transfer function that this system is of second order with double roots on $s=0$ at the denominator. It also means that this open-loop system has double pole which lie right on the imaginary axis of s-plane, implying an unstable nature. It makes sense since the ball always keeps rolling down due to the gravity and never rolls back up to reach set points whenever the beam is steadily tilted up. MatLab simulation shows that the step response blows up as the time goes on, as shown in Figure 2.

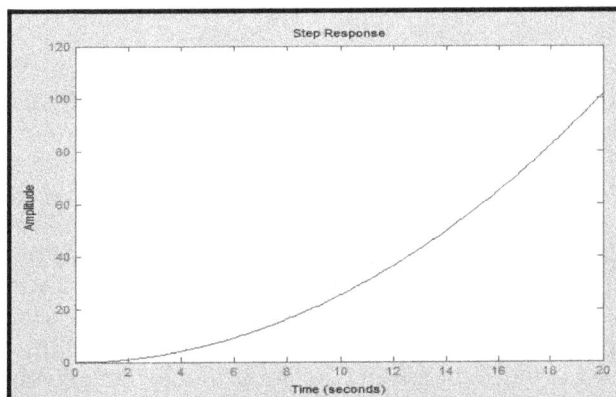

Figure 2. Step response of the open-loop system

3. Design of Hardware and Software

The design of the Ball and Beam consists of the closed-loop control system modeling, mechanical and electronic hardware, and microcontroller software. The overview of the system is shown in Figure 3.

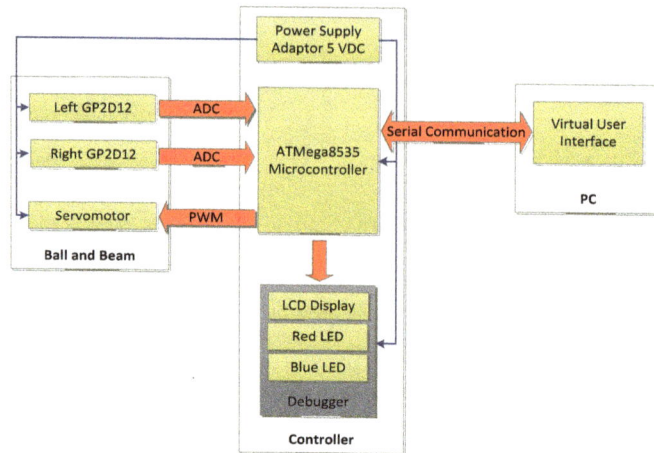

Figure 3. System architecture overview

3.1. Mechanical Design

The designed ball and beam system has a dimension of 90 cm, 15 cm, and 26 cm in length, width, and height, respectively. A billiard ball, having diameter of 5.65 cm and weighing about 180 gram, rolls back and forth by the use of servo-mechanism to reach any desired set points. The mechanical hardware is shown in Figure 4.

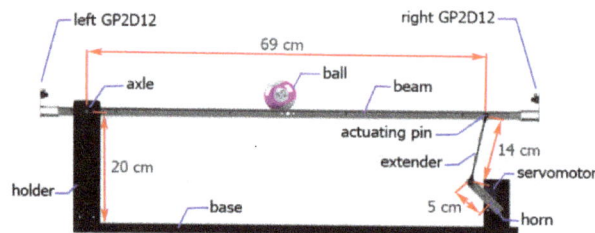

Figure 4. Front view of the mechanical design

Two infrared ranging sensors GP2D12 are placed at both ends of the beam and 5-volt-powered to measure the displacement of the ball. These sensors apply triangulation principle with infrared LED as the emitting element and Position Sensitive Detector as the receiver. An integrated signal processing was pre-embedded in the sensors to produce analog voltages as the output. These sensors are able to detect any objects 10 to 80 cm away. The output voltage of GP2D12 when detecting 10 cm and 80 cm-away objects are 2.6 and 0.4 volts, respectively. The object must be 6 cm in diameter to be properly detected from the distance of 80 cm [11]. In order to acquire valid ball displacement data, the microcontroller sample left and right GP2D12 sensor values with interval of 1 cm. The acquired data are then put into a look-up table, which in turn, build a linear interpolation to determine all possible positions of the ball. The left GP2D12 measures the ball displacement starting from 0 cm to 35 cm, whereas the right GP2D12 measures the ball displacement starting from 35 cm to 60 cm.

The actuator activating the Ball and Beam is a servomotor Turnigy 620MG which is equipped with internal closed feedback inside [12]. A servomotor consists of DC motor, gearbox, variable resistor, and control circuit. The rotational angle of servomotor is determined by the duty cycle of pulses with period of 20 ms. Horn and extender were assembled at the servomotor axle to adjust its angle and thus tilt the beam up and down. A 5 cm horn was meant to increase the angular velocity of the beam. An *R-C* (resistor-capacitor) low pass filter which has cut-off frequency f_c=1.608 Hz was implemented by the means of software to stabilize data acquisition of both left and right GP2D12. Suppose sampling period = Δt, then input voltage Vin(t) was sequentially represented as x_i, x_{i+1}, ..., x_n and output voltage Vout(t) was sequentially represented as y_i, y_{i+1}, ..., y_n, then the discrete low pass filter is given by:

$$y_i = ay_{i-1} + (1-a)x_i \qquad (11)$$

Where *a=RC/(RC+Δt)*

In order to monitor the status of the system, a set of debugger was added into system. A red Blue LED indicated the beam status whether "Run" or "Stop". A Red LED and buzzer indicated whether or not a ball was already on the beam. A 16x2 character LCD displayed information regarding to the serial connection status with user interface program as well as the beam status. The heart of the system is the AVR ATMega 8535 manufactured by ATMEL Corporation [13]. This handy chip facilitates internal Analog-Digital Converter, large program and data memory, internal comparator circuit, internal Pulse-Width Modulation, and internal timer. Subroutines of the controller handle time sampling tasks, data acquisition from GP2D12 sensor, data-filtering using low pass filter, data conversion to displacement, PID calculation, servomotor angle adjustment, and connectivity to user interface program via serial communication on PC or laptop for both sending data to be plotted by user interface program and receiving commands from user interface program, as shown in Figure 5.

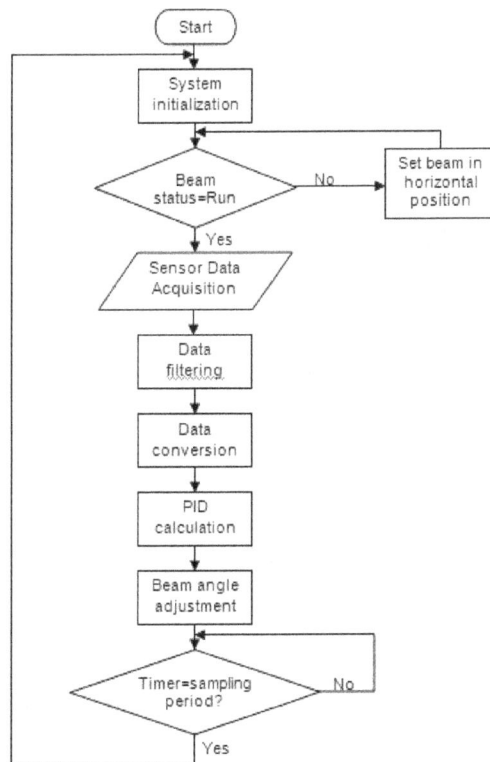

Figure 5. Flow chart of controller subroutines

Figure 6 shows the closed-loop control system consisting of a PID controller, plant (Ball and Beam), and low pass filter. *C(s)* is the transfer function of PID controller in *s*-domain, *G(s)* is the transfer function of the plant, i.e. Ball and Beam system, and *H(s)* is the transfer function of GP2D12 low pass filter. The PID controller was supposed to make the plant stable by shifting the open-loop poles, which originally lie on the imaginary axis, to the left-half s-plane. PID calculation was initialized with error calculation, i.e. difference between set points and process variables. The obtained error was the used to calculate manipulated variable (MV), which is formulated by:

$$MV = K_p * \text{error} + K_i * \text{sum of error} + K_d \left(\text{error} - \text{last error} \right) \qquad (12)$$

Then, the value of MV was used to control servomotor pulse duty cycle through pulse width modulation (PWM). During calculating PID, the integral and derivative sampling period was 50 ms. The transfer function of closed-loop system is now of fourth order and given by:

$$\frac{X(s)}{R(s)} = \frac{0.243 K_d s^3 + 2.45 \left(K_d + 0.099 K_p \right) s^2 + 2.45 \left(K_p + 0.099 K_i \right) s + 2.45 K_i}{0.478 s^4 + 4.83 s^3 + 2.45 K_d s^2 + 2.45 K_p s + 2.45 K_i} \qquad (13)$$

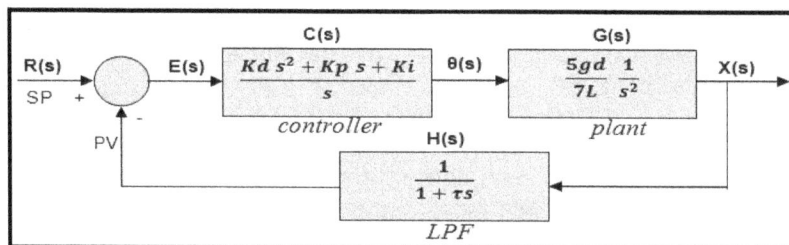

Figure 6. Block diagram of closed-loop system

The realized Ball and Beam system is depicted in Figure 7.

Figure 7. The realized Ball and Beam

3.2. User Interface

The user interface program was used to adjust PID configuration within closed-loop control system, select the desired set points, display the plot of ball displacement and set points, and calculate the rise time, peak time, maximum overshoot, and settling time, as shown in Figure 8. The application design was accomplished in Visual Studio 2010 with Visual Basic programming language (VB.Net). On the user interface main page, there are three major displays, which are: control panel, plotter display, and info display.

a) Control panel consists of four panels: Beam Status, Plot Data, PID Configuration, and Ball Position. At ball position option, users are allowed to select the desired set points, as shown in Figure 9.

b) Plot display has a grid area to plot the current ball position. The vertical axis represents the displacement (cm) ranging from -10 to 70 cm, while the horizontal axis represents the time (s). The yellow line indicates the midpoint of the beam. The purple line indicates the ball displacement, while red line represents a marker on which the response starts settling.

c) Info display shows time response characteristics, including: current position, set point, current error, rise time, maximum overshoot, peak time, and settling time.

While communicating with the microcontroller, an updater program handles data-sending task to user interface program as well as data-receiving from user interface program via serial communication with baud rate of 115200 bps. Along the communication, character string packets were transmitted and then translated into instructions or commands. The format of the character string packets received by microcontroller was defined as #[instruction]X. The character "#" was used as the initial mark while the character "X" was used to end the data packet. The instruction itself lies between character "#" and X. Meanwhile, the microcontroller sent character string packet which no longer needs initial marks but only end marks "/n".

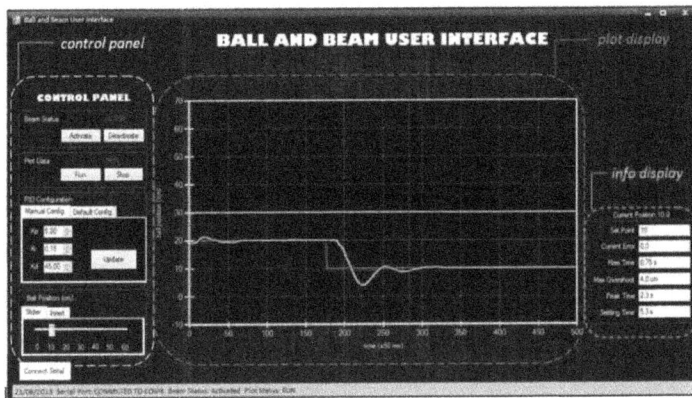

Figure 8. User Interface display

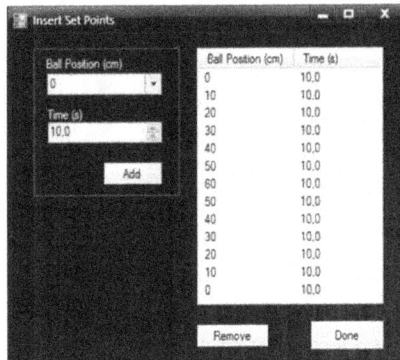

Figure 9. *Insert Set Points* page

4. Results

A set of tests was carried know to assess the overall performance of the Ball and Beam training kit.

4.1. User Interface

Such tests were done to obtain the optimum K_p, K_i, and K_d constants with respect to the rise time, maximum overshoot, peak time, and settling time [14-15], as well as to evaluate the performance of the closed-loop system. Testing mechanisms were carried by adjusting set points from 20 cm to 10 cm, as indicated in Figure 10.

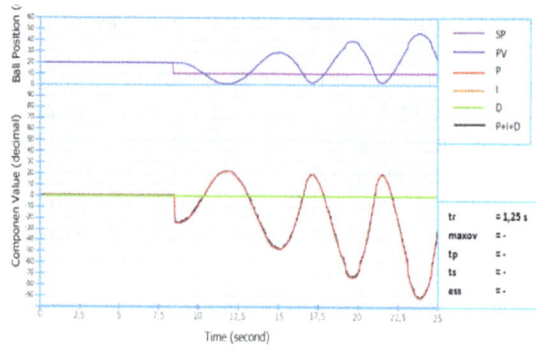

(a)

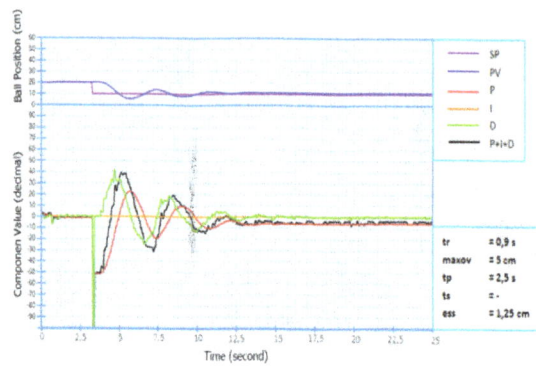

(b)

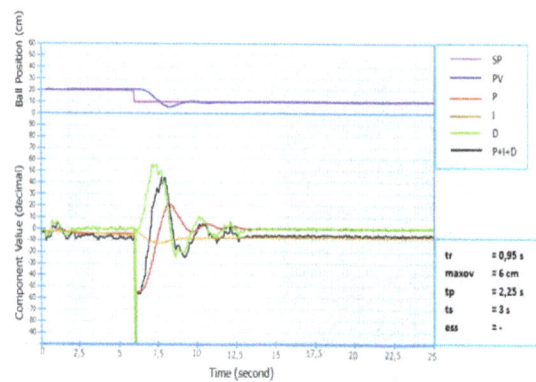

(c)

Figure 10. System performance using (a) Kp=2.5, Ki=0, Kd=0, (b) Kp=5, Ki=0,1, Kd=30, (c) Kp=5, Ki=0,1, Kd=45

    The overall performance of the system needs to be evaluated by calculating rise time, maximum overshoot, peak time, and settling time at all set points using specified K_p, K_i, and K_d configurations. Evaluations were carried by adjusting set points every 15 seconds starting from 0 cm, 10 cm, 20 cm, 30 cm, 40 cm, 50 cm, 60 cm, 50 cm, 40 cm, 30 cm, 20 cm, 10 cm, and finally back to 0 cm. All test configurations were conducted four times and then evaluate the average of rise time, average of maximum peak (Mp), average of peak time, average of settling time, and percentage of successful settling.

4.2. Feedbacks from Students

Seventeen students taking the Control System course were required to fill out a questionnaire as a basic assessment about the use of the microcontroller-based Ball and Beam. The statement in the questionnaire had to be answered on a five-point Likert scale with points 1, 2, 3, 4, and 5 representing "strongly disagree", "disagree", "neutral", "agree", and "strongly agree", respectively. After exploring and practicing the Ball and Beam system, they gave their opinion on the following statements: The Ball and Beam helps students to understand closed-loop control system concepts and is easy to perform. Figure 11 shows the questionnaire result of the statement asked to the students. The result indicates that the microcontroller-based Ball and Beam system is an advantageous tool in achieving better understanding of closed-loop control system concepts and very easy to perform. The result of questionnaire was then converted to Rating Value of 94.12% indicating that the students were very satisfied.

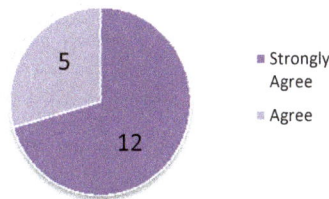

Figure 11. Pie chart of the questionnaire

5. Conclusion

We have developed an ATMega 8535 microcontroller-based Ball and Beam system which consists of infrared ranging sensors, servomotor, PID controller, ball and beam. A virtual user interface, which runs on PC or notebook connected to microcontroller via serial communications, was used to bridge out the system and students. Such user interface was aimed to configure controller coefficients Kp, Ki, and Kd, to select the set points, to plot the actual ball position, and to display parameter values, namely, actual error, rise time, maximum overshoot, peak time and settling time of the system. The PID constants Kp=5, Ki=0.2, Kd=45 was chosen to be the default configuration since it has the largest percentage of successful settling, i.e. 93.75%. This Ball and Beam training tool shows conformities with control system theory, in terms of physical modeling, system stability, and closed-loop PID-controlled system. Feedback from the students indicates that the Ball and Beam system is a helpful tool in achieving better understanding of closed-loop control system concepts and very easy to perform.

References

[1] J Sanchez, S Dormido, F Esquembre. The learning of control concepts using interactive tools. *Comput. Appl. Eng. Educ.* 2005; 13(1): 84-98.

[2] IH Altas, H Aydar, A real-time computer-controlled simulator: For control systems. *Comput. Appl. Eng. Educ.* 2008; 16(2): 115-126.

[3] S Dormido, S Dormido-Canto, R Dormido, J Sanchez, N Duro. The role of interactivity in control learning. *Int. J. Engng. Ed.* 2005; 21(6): 1122-1133.

[4] JL Guzman, KJ Astrom, S Dormido, T Hagglund, Y Piguet. Interactive learning modules for PID control. *IEEE Contr. Syst. Mag.* 2008; 28(5): 118-134.

[5] N Aliane. Spreadsheet-based interactive modules for control education. *Comput. Appl. Eng. Educ.* 2010; 18(1): 166-174.

[6] B Chen, YC Chou, HH Cheng. Open source Ch control system toolkit and web-based control system design for teaching automatic control of linear time-invariant systems. *Comput. Appl. Eng. Educ.* 2013; 21(1): 95-112.

[7] SG Nieto, M Martinez, J Salcedo, D Lauri. Practice tool based on open source Scada for experimentation in nonlinear control using the inverted pendulum. *Comput. Appl. Eng. Educ.* 2012; 20(1): 137-148.

[8] D Ibrahim. Teaching digital control using a low-cost microcontroller- based temperature control kit. *Int. J. Elect. Eng. Educ.* 2003; 40(3): 175-187.

[9] O Gonzalez, M Rodriguez, A Ayala, J Hernandez, S Rodriguez. Application of PICs and microcontrollers in the measurement and control of parameters in industry. *Int. J. Elect. Eng. Educ.* 2004; 41(3): 265-274.

[10] D Ibrahim, T Alshanableh. An undergraduate fuzzy logic control lab using a line following robot. *Comput. Appl. Eng. Educ.* 2011; 19(4): 639-646.

[11] Sharp   Corp.,   GP2D12   Optoelectronic   Device   Datasheet,   2005, http://www.sharpsma.com/webfm_send/1203 [accessed on September 26, 2014].

[12] Hobby   King,   Turnigy   620DMG+HS   High   Torque   Digital   Servo   (MG)   Datasheet, "http://www.hobbyking.com/hobbyking/store/__9441__Turnigy_620DMG_HS_High_Torque_Digital_S ervo_MG_10_6kg_13sec_52g.html [accessed on September 15, 2014].

[13] Atmel Corporation, AVR ATMega 8535 Datasheet, 2006, www.atmel.com [accessed on September 12, 2014]

[14] CL Phillips, RD Harbor. Feedback Control Systems. Second Edition. Englewood Cliffs, New Jersey: Prentice Hall. 1991: 124.

[15] NS Nise. Control Systems Engineering. Fourth Edition. New Jersey: John Wiley & Sons. 2004: 193.

8

Simulation Time and Energy Test for Topology Construction Protocol in Wireless Sensor Networks

Satyam Gupta, Gunjan Gupta

Department of Electronics and Communication Engineering, Invertis University, Bareilly, India
e-mail: satyam9598@gmail.com

Abstract

Coverage area and energy consumption are very big challenges in the field of Wireless Sensor Networks (WSNs) as it affects the number of sensors, connectivity and network. Since the sensors are operating on battery of limited power, it is a challenging aim to design an energy efficient Topology Control protocol, which can minimize the energy and thereby extend the lifetime of the network. Through this paper an attempt has been made in terms of simulation time and spent energy in construction of topology in the sensor network by comparing Just Tree and K-neigh Tree protocols. The result shows that K-neigh Tree protocol consumes less energy than Just Tree protocol.

Keywords: *just tree, K-neigh tree, energy consumption, topology construction, wireless sensor networks (WSNs)*

1. Introduction

Past several years have witnessed a great success of Wireless Sensor Networks (WSNs). As an emerging and promising technology, WSNs have been widely used in a variety of long term and critical applications including event detection, target tracking, monitoring and localization and so on. A sensor network usually consists of n number of sensor nodes which are self organized into multi- hop fashion. By working together, sensor nodes coordinate to finish a common task [1, 7]. Since sensor nodes are very small in size, due to their tiny size, sensor node cannot be equipped with big batteries. Therefore, energy conservation is very crucial in the context of sensor networks. So to conserve energy in sensor networks, one of the approaches proposed so far is Topology Control. The basic idea behind Topology Control is to restrict the network topology, i.e. energy need for transmission depends on the distance between transmitter and receiver. There is wide range of operations that Topology Control can perform to conserve energy for sensor nodes, it includes reducing the transmission range of the nodes and turning off the nodes which are not in use.

The simulation is performed on ATARRAYA software [2] for the design and evaluation of Topology Control algorithms in Wireless Sensor Networks. In Atarraya Software, there are number of Topology Control protocols [2, 6] present which include protocols like A3, A3 Coverage, Simple Tree, Just Tree, K-neigh Tree, EECDS, CDS Rule K. In this work we had compared Just Tree and K-neigh Tree protocols.

Just Tree: The Just tree algorithm [6] assumes one sink node responsible for message/information broadcast. The sink nodes are capable of sending or receiving messages from other neighboring sensor nodes. The message or number of events are propagated within the network using the same concept of parent node and child node, the parent node initiates the message and transfer this message to other sensing nodes acting as child node. The concept of Just Tree ensures that as the deployment area will increases or if the deployment area is constant the number of nodes if increased will denote the increase in the size of the tree in order to efficiently cover a flexible or constant deployment area.

K-neigh Tree: The sink node transmits a HELLO message to all its neighbors at highest power, which contains its ID number and its tree level. The node that accepts the HELLO message stores the ID of transmitted node and tree level, calculates the distance with the transmitting node and sets its state to reside. After accepting the HELLO message for the first time, a node transmits a HELLO message of its own to its own neighbors and sets a timer in order to listen for its neighbor's messages. The K-neigh Tree protocol assumes that the nodes have no knowledge of their locations, and that they can change their transmission power in a regular manner. Using a discrete number of power levels will be studied in a future work [4] [6]. The computational complexity of the protocol depends directly on the selected sorting algorithm, while the message complexity is of 2 messages per node, which could be reduced to 1 message if the UPDATE message is not sent.

2. Simulation Results

The algorithms mentioned above are evaluated on simulator ATARAYA which was specifically designed for Wireless Sensor Networks. The Simulator allows us to select the different network parameters, such as simulation time, number of message transferred (events), Queue size and Energy Consumed.

3. Analysis

The result of simulation produced in Table 1 and 2 shows that all the algorithms have their individual and independent role for construction of topology. The protocols were evaluated on a specifically designed simulator for WSN topology. The simulator Atarraya allows the scalability of the underlying network with the ease of selecting different network parameters, such as deployment area, number of nodes and communication range.

Table 1. Performance of Just Tree Algorithm

Nodes	Simulation Time (sec)	Consumed Energy (mJ)
100	13.420	22.917
150	9.337	45.768
200	8.092	72.714
250	7.725	101.597
300	7.087	155.007
350	7.113	191.539
400	6.020	237.844
450	6.408	289.575
500	5.446	367.961

Table 2. Performance of K-neigh Tree Algorithm

Nodes	Simulation Time (sec)	Consumed Energy (mJ)
100	4.485	22.553
150	4.110	43.836
200	4.250	72.670
250	4.218	101.665
300	4.094	146.147
350	3.894	185.190
400	3.886	240.264
450	3.989	296.363
500	3.926	361.614

As clearly shows in tables that each parameter of every protocol increases/decreases with increase in number of nodes. But when we talk about the comparative study of the two protocols then we find that simulating results produced by K-neigh Tree protocol had given the best results among the other Topology Control protocol as it takes less simulation time and less amount of energy spent in Topology Control in comparison with Just Tree protocol.

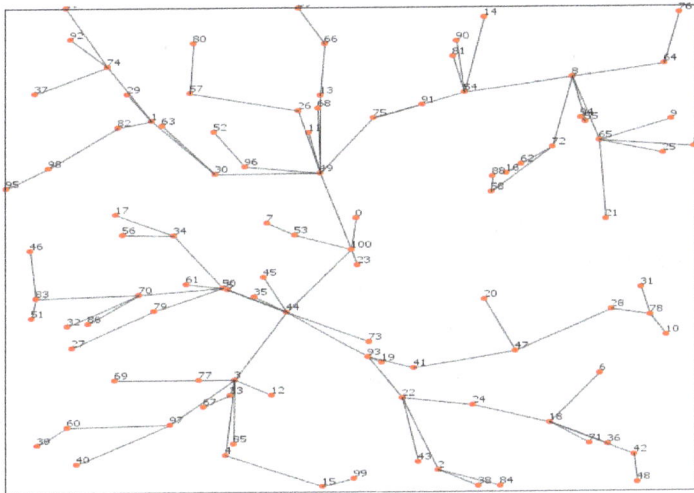

Figure 1. Deployment of 100 nodes

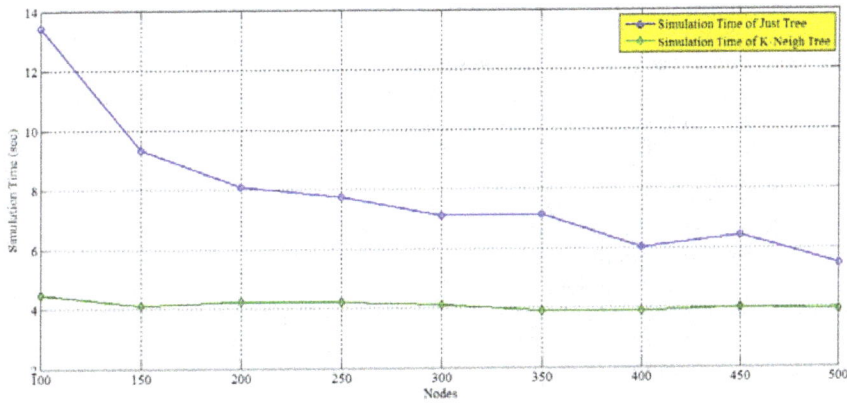

Figure 2. Nodes v/s Simulation time of Just Tree and K-neigh Tree

In Figure 2, as we clearly observes that simulation time taken by K-neigh Tree protocol is almost constant while simulation time of Just Tree goes on decreasing and taking more time to control topology than K-Neigh Tree protocol.

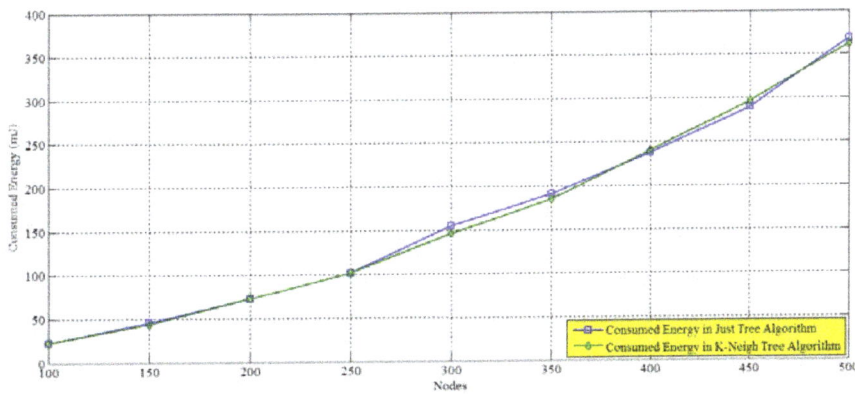

Figure 3. Nodes v/s Consumed Energy in construction of Just Tree and K-neigh Tree

In Figure 3, we are talking about Consumed Energy in both topology control algorithms and we observe that K-neigh Tree algorithm has consumed less energy although there is a minute difference between the energy consumed in K-neigh Tree and Just Tree algorithm, i.e. of 6.347mJ.

4. Conclusion

The above given graphs clearly shows that how differently these Topology Control algorithms work. That Topology would be beneficial for us that took less time and spent less amount of energy so that our sensors may work for long lasting.

Through the comparative study of the two algorithms, i.e. Just Tree and K-neigh Tree, we found that the simulation time for Just Tree algorithm remains almost constant for 500 nodes and spent energy level also is more than that of K-neigh Tree algorithm. So on the basis of above obtained graphs for 500 nodes, we can conclude that K-neigh Tree algorithm is better than Just Tree algorithm as it takes less time and utilizes less amount of energy spent in Topology Construction. Further in future lot of work would be carried out in other topologies as mentioned above.

References

[1] Pedro Mario Wightman R, Miguel A Labrador. Reducing the communication range or turning nodes off? An initial evaluation of topology control strategies for wireless sensor networks. Ingeniería & Desarrollo. Universidad Del Norte. 2010; 28: 66-88.
[2] Pedro M Wightman. Atarraya: A Simulation Tool to Teach and Research Topology Control Algorithms for Wireless Sensor Networks. Simulation tool. 2009.
[3] Pedro M Wightman, Miguel A Labrador. A3: A Topology Construction Algorithm for Wireless Sensor Networks. *IEEE*. 2008.
[4] A Karthikeyan, T Shanker, V Srividhya, Siva Charan Reddy V, Sandeep Kommineni. Topology Control Algorithm for Better Sensing Coverage with Connectivity in Wireless Sensor Networks. *Journal of Theoretical and Applied Information Technology*. 2013; 52(3).
[5] P Santi. Topology Control in Wireless Adhoc And Sensor Networks. John Wiley and Sons. 2005.
[6] PM Wightman, MA Labrador. Topology Control in Wireless Sensor Networks. Springer: 2009.
[7] Tarun Dubey, OP Sahu. Survey on Wireless Sensor Networks for Reliable Life Services and Other Advanced Applications. *Indonesian Journal of Electrical Engineering and Informatics*. 2013; 1(4): 133-139.

Modelling of a Trust and Reputation Model in Wireless Networks

9

Modelling of a Trust and Reputation Model in Wireless Networks

Saurabh Mishra
Department of Electronics and Communication Engineering
Invertis University, Bareilly, Uttar Pradesh 243123, India
Email: saurabhmishra.er@gmail.com

Abstract

Security is the major challenge for Wireless Sensor Networks (WSNs). The sensor nodes are deployed in non controlled environment, facing the danger of information leakage, adversary attacks and other threats. Trust and Reputation models are solutions for this problem and to identify malicious, selfish and compromised nodes. This paper aims to evaluate varying collusion effect with respect to static (SW), dynamic (DW), static with collusion (SWC), dynamic with collusion (DWC) and oscillating wireless sensor networks to derive the joint resultant of Eigen Trust Model. An attempt has been made for the same by comparing aforementioned networks that are purely dedicated to protect the WSNs from adversary attacks and maintain the security issues. The comparison has been made with respect to accuracy and path length and founded that, collusion for wireless sensor networks seems intractable with the static and dynamic WSNs when varied with specified number of fraudulent nodes in the scenario. Additionally, it consumes more energy and resources in oscillating and collusive environments.

Keywords: *WSNs, static wireless (SW), dynamic wireless (DW), trust and reputation models (TRMs), malicious nodes.*

1. Introduction

Past several years have witnessed a great success of wireless sensor networks (WSNs). As an emerging and promising technology, WSNs have been widely used in a variety of long term and critical applications including event detection, target tracking, monitoring, and localization. In recent years, the basic ideas of trust and reputation have been applied to WSNs to monitor the changing behaviors of nodes in a network. Several trust and reputation monitoring (TRM) systems have been proposed, to integrate the concepts of trust in networks as an additional security measure, and various surveys are conducted on the aforementioned system. However, the existing surveys lack a comprehensive discussion on trust application specific to the WSNs. This survey attempts to provide a thorough understanding of trust and reputation as well as their applications in the context of WSNs. The survey discusses the components required to build a TRM and the trust computation phases explained with a study of various security attacks. The sensor networks are constructed by a large number of nodes with ultra-low power computation and communication units [1]. An adversary can control a sensor node undetectably by physically compromising the node and use the captured nodes to inject faulty or false data into the network system disturbing the normal cooperation among nodes. Authentication and cryptographic mechanisms alone cannot be used to full solve this problem because internal adversarial nodes will have valid cryptographic keys to access the other nodes of the networks. Many existing approaches at most concentrate on cryptography to improve data authentication and integrity but this addresses only a part of the security problem without consideration for high energy consumption. Monitoring behavior of node neighbors using reputation and trust models improves the security of WSNs and maximizes the lifetime for it. However, a few of previous studies take into consideration security threats and energy consumption at the same time. WSNs serve to gather data and to monitor and detect events by providing coverage and message forwarding to base station. However, the inherent characteristics of a sensor network limit its performance and sensor nodes are supposed to be low-cost. An attacker can control a sensor node undetectably by physically exposing the node and an adversary can potentially insert faulty data or misbehavior to deceive the WSNs. Authentication mechanisms and cryptographic methods alone cannot be used to completely solve this problem because internal malicious nodes will have valid cryptographic keys to access the other nodes of the networks. Also conventional security methods cannot be used for WSNs due to power and processing limitations. Recently, a new mechanism has been offered for WSNs security improvement. This mechanism relies on constructing trust systems through analysis of nodes observation about other nodes in the network [2, 3]. A sensor node is always at risk of being compromised by an adversary, who may capture the node's cryptographic keys. Such an attack is also referred as insider attack [5] in which an adversary node would appear to be a legitimate member of the network. Once a sensor node is captured, an adversary may sniff and inject packets with falsified data that may compromise the node's data integrity. Therefore, security and privacy challenges of WSN must be addressed to prevent the system from turning against those for whom the system has to render benefit. Although external security attacks on WSN may be countered by the use of cryptographic techniques, cryptography is not that effective against the internal insider attacks by the malicious node. One approach that has gained global recognition in providing an additional means of security for decision making in WSNs (i.e., to trust a node for communication or not) is the trust and reputation monitoring (TRM) system. TRM deals with the problem of uncertainty in decision making, by keeping the history of a node's previous behavior (repute). A node is trusted and will be forwarded with the packets only if the node holds a good repute; otherwise, the node will be considered untrustworthy. TRM provides a natural choice for security in open systems, the Internet and social networking for being computationally tractable.

2. Background of Trust and Reputation Models

This section reports surveys of five trust and reputation models briefly with the work on wireless sensor networks.

Eigen Trust Model: It is one of the most commonly used trust and reputation models in the wireless sensor network domain. Kamvar et al. [4] evaluated this model on the basis of the peer's history of contributions by assigning a unique global trust value in the peer-to-peer file system for each peer [5, 6]. Further into this model, the authors define Sij as the local trust of peer i about peer j, in the following Equation (1):

$$Sij = \text{sat}\ (i, j) - \text{unsat}\ (i, j) \tag{1}$$

Equation (1) shows the difference between satisfactory and unsatisfactory interaction between peers: (i, j). Further, the authors define normalized local trust value in Equation (2):

$$Cij = \max (Sij, 0)/\Sigma j \max (Sij, 0) \tag{2}$$

The above equation ensures that all the value lies in between 0 and 1.

Peer Trust Model: In this model many aspects related to the trust and reputation management such as the feedback a peer receives from other peers, the total number of transactions of a peer, the credibility of the recommendations given by a peer, the transaction context factor and the community context factor are combined.

Bio-inspired Trust and Reputation Model (BTRM-WSN): This model for wireless sensor networks is based on the bio-inspired algorithm of ant colony system. In this model, most trustworthy path leads to finding the most reputable service provider in a network.WSN launches a set of artificial agents while searching for a most reputable service provider.

LFTM Model: This linguistic fuzzy trust model uses the concept of fuzzy reasoning. On one hand, it uses the representation power of linguistically labeled as fuzzy sets for the satisfaction of a client or the goodness of a server. On the other hand, it remains affected by the inference power of fuzzy logic, as in the imprecise dependencies between the originally requested service and the actual received one, or the punishment to apply in case of fraud. The expected result will be an easily interpretable system with adequate performance. In this model, a set of linguistic labels describing several levels of a variable or concept could be associated with a fuzzy set. The resultant set constitutes linguistic labels such as VERY LOW, LOW, MEDIUM, HIGH and VERY HIGH. These defined fuzzy sets associated with such labels specify the level of client satisfaction.

3.   Modified Trust and Reputation Models

To choose accurate trust and reputation models remains the top priority for the performance assessment of wireless sensor networks. Optimal trust and reputation models enhance the performance of the overall system about information dissemination, but the wireless sensor network system may not be dependent on the same. A simple trust and reputation modeling strategy may give the best result for a single instance but we have to deploy such efficient trust and reputation modeling strategies that provide optimal results in data dissemination. The improper modeling strategy may overload the entire network and consume more resources both in terms of energy and computation which result in the entire system performance degradation [14]. There always remains dire influence of trust and reputation strategy on the entire operating environment when evaluating a specific wireless sensor network. The goal which remains there are to carefully choose and examine the trust and reputation modeling strategies for information dissemination and present an optimal result without compromising any constraints than the expected outcome. Therefore, a typical realization should be required to access the scope of a particular trust and reputation model strategy for the wireless sensor networks. In our analysis, we consider ten networks composed of two hundred sensor nodes, each for twenty scenarios in two-dimensional fields. Sensor nodes in a cluster with a specific radio range transmit the data to the cluster head and then to the base station within the entire network. Network deployment focuses on collusion and oscillating conditions.

Although any trust and reputation sensor node strategy can be used in our model, we used Eigen trust model with static, dynamic and oscillating wireless sensor network for our proposed framework. Static wireless sensor network can be referred to as a mode of communication where the position of all the nodes remains stationary, whereas in case of dynamic wireless sensor network, the nodes can change their positions in an accord manner. Accordingly, for a given network with static and dynamic wireless sensor network and trust and reputation models node strategy described above, we are interested in finding the following two problems: (i) what is the influence of variation of collusion on static and dynamic communication node operations in the wireless sensor networks and (ii) how the varying collusive environment affects the accuracy and path length for different modes of Eigen trust and reputation model in wireless sensor network.

3.1. Simulation for Related Research

Although there is much research work conducted in application of trust and reputation in various network domains, the task is still in evolutionary phases in the case of WSNs, where node security is the biggest challenge because of low resources of node. Every proposed TRM has some limitations and covers only a subset of various issues and challenges in providing complete security to a WSN. Some of the issues for future research can be considered in the field of TRMs, which are discussed in the subsequent paragraphs.

In a few TRMs, to survive in the network, a node must continuously contribute to the network traffic. Nodes in the low activity areas of a network may suffer because of their gradual decrease in reputation. Therefore, a mechanism must be devised to keep the repute above a threshold in such low activity areas.

In most of the trust models, a node calculates the direct trust through promiscuous learning mode. However, when directional antennas are used, the technique becomes difficult to implement. Similarly, noise may be another factor that can cause hindrances to watchdog mechanisms [14].

Most of the TRMs that we discussed in this survey use a flooding approach for trust information dissemination, and this may lead to high traffic over the network. With the addition of more nodes in the network, the performance may further degrade. Therefore, an optimal trust and reputation modeling strategy may give the best result for a single instance but we have to deploy such efficient trust and reputation modeling strategies that provide optimal results in data dissemination. The improper modeling strategy may overload the entire network and consume more resources both in terms of energy and computation which result in the entire system performance degradation.

We focused on two parametric aspects, namely: accuracy and path length for information dissemination in wireless sensor networks. For this, we have developed the unmitigated scenario pinpointing two main targets. Firstly, we are interested in finding the value of two above-mentioned parameters for static and dynamic wireless sensor network with and without collusion aspect. We want to know the summation of all the node operations with respect to collusion parameter. Lesser path length of node operation always gives due attention as it consumes fewer resources and exhibits more efficiency. Secondly, we want to make an estimation of the mobility effects on communication performance in correlation with the collusion for Eigen trust and reputation model. We designed a wireless sensor network template using the following parameters: 20% of all nodes in a randomly created WSN acted as clients where and the rest 80%of nodes acted as servers. Client nodes refer to the percentage of nodes which want to have or ask for services in a WSN. 5% of the nodes acted as relay servers which do not offer any services and act as relay nodes. The radio range of the nodes set at 10 hops to its neighbors. We consider a scenario where the percentage of fraudulent servers varied from 10% to 100% which specifies the indispensable condition for our WSN framework evaluation. Fraudulent servers depict the percentage of adversaries in a wireless sensor network. We set the minimum and maximum numbers of nodes that can create a WSN equal to 200.

Sensor nodes belonging to our developed networks spread over the area of 100m × 100 m. A total of ten networks were examined and the final results reflect the average value of all the networks. The process of searching trustworthy server was carried out ten times for each network.

3.2. Simulation Results

This section enables us to implement and evaluate Eigen trust and reputation model for different wireless sensor network modes. We used Java based event driven TRMSim-WSN simulator [9] version 0.5 for wireless sensor network allowing the researchers to simulate and represent random network distributions and provides statistics of different data dissemination policies including the provision to test the different trust and reputation models' strategies. Many decisions, like static or dynamic or oscillating networks, a combination of dynamic and oscillatory networks, the percentage of fraudulent nodes, the percentage of nodes acting as clients or servers, and so forth, can be implemented and tested over it. The proposed model is tested on five different modes with varying fraudulent conditions. We reported a comprehensive analysis based on collusion with static and dynamic wireless sensor networks.

We collected data for two metrics, namely, accuracy and path length. We investigated the comparative analysis of different modes with oscillating, static WSN and dynamic in contrast with and without collusion parameter. Static node refers to the type of nodes whose position remains fixed and whereas the dynamic node can be mobile in the network. We considered five WSN modes, namely, (i) Static WSN (SW) (ii) Static WSN with collusion (SWC) (iii) Dynamic WSN (DW) (iv) Dynamic WSN with collusion (DWC) (v) Oscillating WSN.

Accuracy: The term accuracy in the trust and reputation systems may be defined as the selected percentage of trustworthy nodes. We calculated accuracy parameter in terms of their current and average values. Current accuracy denotes the trustworthiness value calculated for the last node, whereas average accuracy presents the value of all nodes available in the mentioned framework. We calculated current and average accuracy corresponds to different WSN modes for Eigen Trust model. According to Figure 1, for 10% of fraudulent environment, the value of current accuracy is highest in case of static WSN as compared to the rest of the WSN modes and lowest in case of dynamic WSN   because of the fact that static nodes are less prone to failure than the dynamic as well as the combination of static and dynamic WSN with collusion aspect. The value of current accuracy remains highest in case of DWC mode as compared to the rest of the WSN modes for 20%, 30%, 40% and 70% of fraudulent environment and lowest in case of dynamic, static and static with collusion WSN for the same environment. For maximum collusive network this value is lowest in case of static WSN and highest for SWC mode.

Next, we considered the second evaluation for average accuracy with the same WSN framework. According to Figure 2, again average accuracy shows that the value of average accuracy remains highest for 10%, 30% and 40% fraudulent environment in case of dynamic WSN than the rest of the WSN modes. For static WSN (SW), this value remains lowest, when malicious server strength is 20%, 30% and 50%. In case of static with collusion (SWC) WSN it is highest in extreme fraudulent environment and lowest when malicious nodes percentage is 40%. The value is highest in case of dynamic WSN with collusion (DWC) mode for 50%, 60% and 70% collusive environment in the Eigen trust model. The oscillating mode outperforms the rest of the modes in average accuracy values. In this mode average accuracy is highest at 80% and lowest at 60% and 70%.

Path Length: The next parameter of our concern is path length which can be defined as the number of resources a particular network utilizes with a particular trust and reputation model. In the consistent pattern of accuracy evaluation types, we evaluated the current and average path length on the similar pattern of accuracy for all the WSN modes. Current path length depicts the resource utilization value calculated for the last node, whereas average path length exhibits the value of all nodes present in the scenario. Figure 3 and 4 represent the value of current and average path length which remains quiet in case of dynamic (DW) WSN and dynamic with collusion (DWC) WSN for both the current and average case viewpoints than other modes. Among the rest of the modes, oscillating and static (SW) mode consumes lesser path length than the rest of the modes in the case of 30% and 80% fraudulent environment for current accuracy, whereas the SWC mode for average accuracy utilizes the minimum path length for 70% fraudulent servers. We also observed that oscillating mode utilizes the maximum path length for 40% to 60% strength of malicious servers.

We proposed a more robust framework subsuming different WSN versus collusion scalability on a single platform.

We enhanced the contribution to a certain extent by incorporating collusion and oscillation parameters for wireless sensor network evaluation making our investigation more robust and real time.

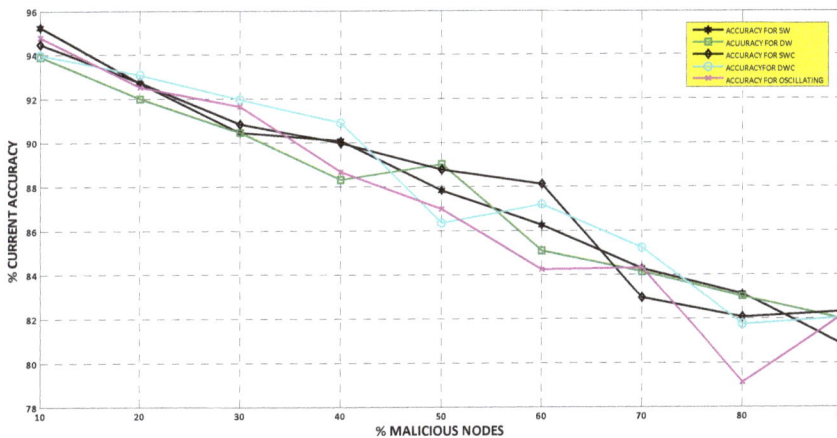

Figure 1. Current accuracy of different WSN modes with varying fraudulent WSN for Eigen trust and reputation model

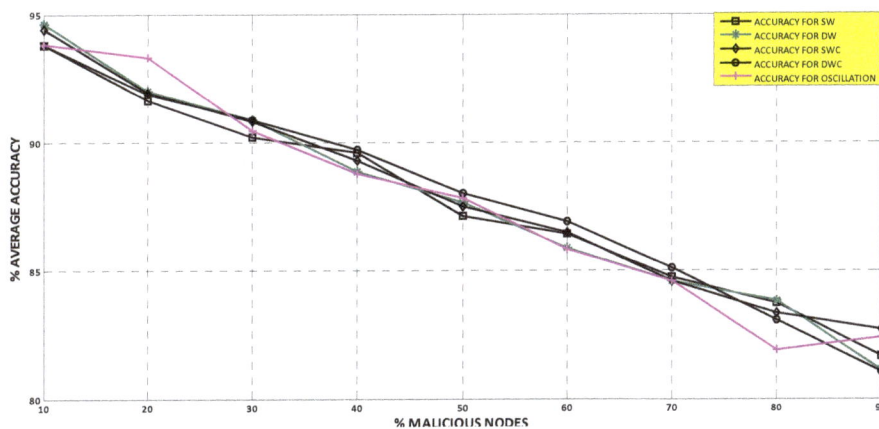

Figure 2. Average accuracy of different WSN modes with varying fraudulent WSN for Eigen trust and reputation model

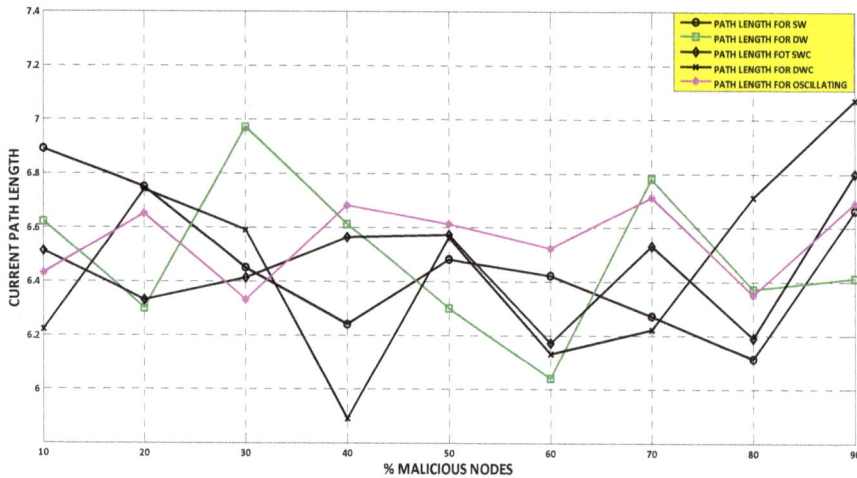

Figure 3. Current path length of different WSN modes with Eigen trust mode

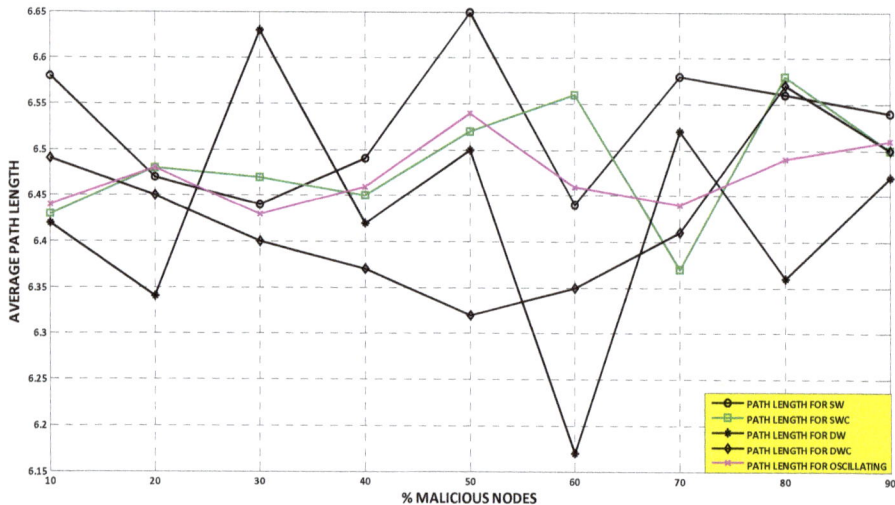

Figure 4. Average path length of different WSN modes with Eigen trust model

4. Conclusion

Trust is an important tool for self-configuring and autonomous systems, such as WSNs, to make effective decisions in detecting a misbehaving node. The task of establishing trust and reputation becomes more challenging when the nodes are mobile. This paper concluded the impact of varying collusion on different trust and reputation modes in wireless sensor networks. We have observed the effect of collusion for static, dynamic, collusive and oscillating sensor nodes in a WSN framework. It is evident from the simulation that there is a strong relationship between collusion and WSN modes in trust and reputation model evaluation. We evaluated a wireless sensor network framework for varying collusion aspect with reference to two performance metrics, namely: accuracy and path length viewpoint. We estimated accuracy and path length in terms of overall percentage of the functionality for sensor node operations. The performance of the WSN system changes along with the different WSN modes and strength of collusion present in the scenario. We mainly concentrated toward the comparative evaluation of static, dynamic, oscillating and collusive WSN modes deployed in our designed model. Our research work presented a comprehensive investigation over collusion parameters with Eigen trust and reputation model. We stressed on two major directions. Firstly, we evaluated accuracy and path length for collusive and non collusive modes of wireless sensor networks. Secondly, we investigated the entire framework for comparative evaluation of above-discussed modes. We observed that with the collusion adoption in the WSN modes, the result becomes much steeper that is, performance degradation. In case of static nodes, the collusion affects less to WSN when it is incorporated in dynamic mode. Also, node operations remain more in case of collusion than without it. From this investigation, we can predict that the lesser the collusive nodes the more the probability of accuracy, the better resource utilization of the entire WSN will be exhibited by the wireless sensor network system.

In the future, we would like to develop further trust and reputation models in our evaluation as well as work towards additions on newer distribution strategies for the wireless sensor network domain.

References

[1] S Farahani. Zig Bee Wireless Networks and Transceivers. Oxford, UK: Elsevier. 2008.

[2] A Alkalbani, T Mantoro, AO Md Tap. *Improving the lifetime of wireless sensor networks based on routing power factors.* In Networked Digital Technologies of Communications in Computer and Information Science. Springer, Berlin, Germany. 2012; 293: 565-576.

[3] H Chen, H Wu, X Zhou, C Gao. *Reputation-based trust in wireless sensor networks.* In Proceedings of the International Conference on Multimedia and Ubiquitous Engineering (MUE'07). Seoul, Republic of Korea. 2007: 603–607.

[4] M Dorigo, L Gambardella, M Birattari, A Martinoli, R Poli, T St¨utzle. Ant Colony Optimization and Swarm Intelligence. *Computer Science.* Springer, Berlin, Germany. 2006; 4150.

[5] O Cord´on, F Herrera, T. St¨utzle. A review on the ant colony optimization meta heuristic: basis, models and new trends. *Mathware & Soft Computing.* 2002; 9(2-3): 141-175.

[6] M Dorigo, T St¨utzle. Ant Colony Optimization. Bradford Book. 2004.

[7] FG M´armol, GM P´erez. Providing trust in wireless sensor networks using a bio-inspired technique. *Telecommunication Systems.* 2011; 46(2): 163-180.

[8] FG M´armol, GM P´erez. TRIP, a trust and reputation infrastructure-based proposal for vehicular ad hoc networks. *Journal of Network and Computer Applications.* 2012; 35(3): 934-941.

[9] FG M´armol, GM P´erez. *TRMSim-WSN, trust and reputation models simulator for wireless sensor networks.* In Proceedings of the IEEE International Conference on Communications (ICC '2009). Dresden, Germany. 2009: 1-5.

[10] VK Verma, S Singh, NP Pathak. Analysis of scalability for AODV routing protocol in wireless sensor networks. *Optik—International Journal for Light and Electron Optics.* 2014; 125(2): 748-750.

[11] AS Alkalbani, AO Md Tap, T Mantoro. *Energy consumption evaluation in trust and reputation models for wireless sensor networks.* In Proceedings of the 5th International Conference on Information and Communication Technology for the Muslim World. Rabat, Morocco. 2013: 1-6.

[12] S Chen, Y Zhang, G Yang. Parameter-estimation based Trust model for unstructured peer-to-peer networks. *IET Communications.* 2011; 5(7): 922–928.

[13] Y Pan, Y Yu, L Yan. An improved trust model based on interactive ant algorithms and its applications in wireless sensor networks. *International Journal of Distributed Sensor Networks.* 2013; 2013: 9.

[14] OP Sahu, Tarun Dubey. A Fault Tolerant Topology for Reliable Data Forwarding in Localized Wireless Sensor Network. *Journal of Instrument Society of India.* 2011; 41(4): 245-247.

10

Economic Selection of Generators for a Wind Farm

Omid Alavi*, Behzad Vatandoust
Department of Electrical Engineering, K.N. Toosi University of Technology, Tehran, Iran
*Corresponding author, e-mail: alavi.omid@mail.com

Abstract
The selection suitable generator for wind turbines will be done based on technical criteria and priorities of the project. In this paper, a method for determining the type of wind turbine generator with an example is explained. In the paper, for a 10kW wind turbine, two generators have been proposed. The first case is a squirrel-cage asynchronous generator coupled to the turbine through the gearbox and directly connected to three phase output. Other PM generators that are directly coupled to the turbine and it is connected to the grid using the inverter. The results show that according to wind conditions, a 10kW permanent magnet generator is more advantageous in terms of energy production.

Keywords: wind turbine, weibull, turbine generator, permanent magnet

1. Introduction

Selection of generator type for a wind farm depends on many parameters, such as access to the power grid, power generator, cost limit, turbine type, quality and quantity of wind and priorities such as efficiency, reliability and the maximum energy. Perhaps In most cases, all of the above are effective in choosing a generator. Usually, according to the priorities of each project, some of these factors are more important. For example, in Reference [1], Selection of turbine generator on the basis of economic aspects of generator stability was evaluated and an objective function based on the mean time to first failure and the mean time between two failures. In Reference [2] costs of a wind turbine are provided separately. Price in different parts of the unit (such as blades, gearbox, generator, etc.) based on the function of the diameter of the turbine is calculated. Therefore the cost function was obtained and Due to the energy taken to try to minimize it. In Reference [3] using the Monte Carlo method type of power generator units can be selected. Power Production capacity and production cost factor can be considered as optimization objectives. In Reference [4], using VSC, producing power generators in a wind farm is transferred to the DC bus and then used another converter for the transfer of power to a three-phase AC grid. Turbines always work at a speed that which have the most power to the grid. So it has not a speed governor. This variable speed is not an issue due to using the inverter. In Reference [5] layout optimization of the turbines in a wind farm is done. With the layout and the proper placement of the turbines overall efficiency has increased about 3%. In Reference [6], optimized for PM synchronous generators with direct coupling to obtain the maximum power from the variable wind speed is done. In Reference [7], using the control of the blade pitch it has tried to obtain the maximum energy at low wind speeds. By controlling the pitch at high wind speed output power of the generator is limited to rated value. In Ref. [8], suitable turbine for a rural with 36500kWh annual consumption energy is selected from different turbine manufacturers. The basis of choice of 30kW turbine for the power supply was the maximum efficiency during the year (the highest capacity factor) and climatic conditions of the region. In this paper, a method for selecting the 10kW generator for wind turbine is provided. The basis of selection of obtaining maximum energy from the generator is respect to costs. For this purpose, the objective function is defined based on the cost-benefit consist of the cost of the generator and incurred equipment, the fixed cost of turbines and other equipment, the energy consumption during a year. To calculate the consumption energy, we used the Weibull curve for Sabzevar city. In this project, optimization carried out by two different generators:

1) Asynchronous generator with a direct connection (Direct Online) to the grid and coupled via the gearbox

2) Axial flux PM synchronous generator connected to the grid through an inverter and a direct coupling.

In other words, the first is a fixed speed and the second is a variable speed type. In any case, considering the costs and the produced energy, optimized power generator is calculated and finally, two generators have been compared in terms of cost and benefit.

2. Characteristics of Wind and Turbine

In every region the distribution of wind follows Weibull function that this can be described by Equation (1):

$$h(v) = \left(\frac{k}{c}\right)\left(\frac{v}{c}\right)^{k-1} e^{-\left(\frac{v}{c}\right)^k} \tag{1}$$

h: Weibull distribution function
c : Scale factor (m/s)
k : Shape factor (m/s)
v : Wind speed (m/s)

In this study, we used the standard deviation method to compute Weibull parameters.
To calculate the coefficient of cumulative and frequency distribution of Weibull, first it is necessary to calculate its first parameter that is k shape.
Using this method, k and c are calculated respectively as [9]:

$$k = \left(\frac{\sigma}{v}\right)^{-1.086} \tag{2}$$

$$c = \frac{v}{\Gamma\left(1+\frac{1}{k}\right)} \tag{3}$$

$$\Gamma(x) = \int_0^\infty exp(-u)u^{x-1}dx \tag{4}$$

In order to calculate the mean wind speed, v, and standard deviation of wind speed, σ, Equation (5), (6) can be used:

$$v = \frac{1}{n}\sum_{i=1}^n v_i \tag{5}$$

$$\sigma = \sqrt{\frac{1}{n-1}\sum_{i=1}^n (v_i - v)^2} \tag{6}$$

In terms of Weibull distribution function, v and σ can be obtained as follows [9]:

$$v = \int_0^\infty v f_w(v)dv = c\ \Gamma\left(1+\frac{1}{k}\right) \tag{7}$$

$$\sigma = \sqrt{c^2[\Gamma(1+2/k) - \Gamma(1+1/k)^2]} \tag{8}$$

Figure 1 shows the Sabzevar Weibull diagram based on c=7.29m/s and k=1.73:

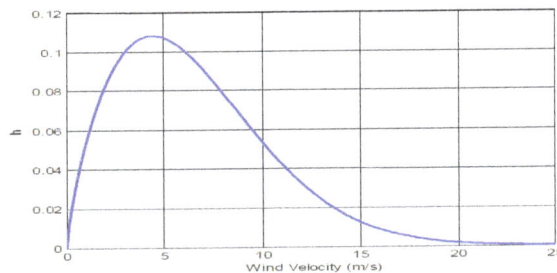

Figure 1. Weibull diagram in Sabzevar region

If we want to calculate the probability that the wind speed is between v_0 and $v_0 + \Delta v$, it is sufficient to take the integral of the function h between these two speeds.

To calculate the wind duration (in hours) between v_0 and $v_0 + \Delta v$ throughout the year, we multiply the probability achieved in a total of 8760 hours of the year:

$$\Delta t = 8760 \times \int_{v_0}^{v_0 + \Delta v} h(v)dv \qquad (9)$$

Table 1 shows the wind speed data in Sabzevar. Using these data, the parameters k and c are calculated for the Weibull diagram.

Table 1. The Speed and Wind Power Density in Sabzavar Region

Month	k	c (m/s)	Average power density (W/m^2)	Most probable speed (m/s)
January	1.72	5.77	187.67	3.48
February	1.68	6.44	271.65	3.75
March	1.75	7.22	361.69	4.43
April	1.67	7.42	419.83	4.29
May	1.70	7.66	446.12	4.55
June	1.74	7.66	446.12	4.55
July	1.71	8.20	542.92	4.92
August	1.71	7.82	470.96	4.68
September	1.72	7.90	480.89	4.76
October	1.75	6.99	326.06	4.31
November	1.73	6.07	217.10	3.70
December	1.67	5.68	188.18	3.28
Annual	1.73	7.29	375.15	4.43

The received power of the wind turbine can be calculated from Equation (10):

$$P_{mech} = \frac{1}{2}C_p A\rho v^3 \times 10^{-3} \qquad (10)$$

P_{mech}: The received power from wind (kW)
C_p : Turbine performance coefficient
ρ : Bulk density of air (1.1 kg/m3 in 1000m)
A : Area swept by the turbine (m2)

C_p is a quadratic function of ratio of the wind turbine speed in behind of the turbine to front of the turbine and ideally, the maximum value is 0.5. Since the wind speed at the behind of the turbine depends on the geometry of the blades, C_p is expressed based on the ratio of the linear velocity of the tip of the blade to the wind speed or the TSR:

$$TSR = \lambda = \frac{r\omega}{v} \qquad (11)$$

r: Radius of the turbine (m)
ω : Angular velocity of the turbine (rad/s)

If the turbine is equipped with a pitch angle control, C_p will be a function of the angle of the blades (β). Turbine studied in this paper is lacking the turbine pitch angle control. Figure 2 shows the variation of C_p in terms of λ:

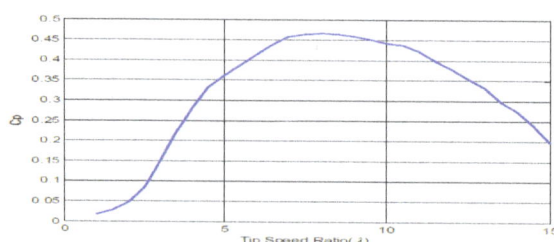

Figure 2. C_p in terms of λ

This diagram is to be drawn by using the design software QBlade for the turbine and its characteristics are as follows:

Diameter turbine=10m

Tower height=11m

v_{cut-in} = 3m/s

$v_{cut-out}$ = 12m/s

3. Energy Calculation

Machine efficiency is usually in nominal operating point. While efficiency is a curve that it is started with zero in non-load and with increasing engine load, increases to a peak (which may be the same as the nominal operating point), and then begins to decline. If we consider a constant value for efficiency from non-load to full load large errors may occur. On the other hand, the machine efficiency diagram is usually not available. For less errors, the machine losses divided into two parts: Constant losses P_{rot} and Load dependent losses P_{vLoss}.

The following values show the different parts efficiency in the rated load:

η_{PM} : PM generator efficiency at full load = 0.92

η_{ind} : Asynchronous generator efficiency at full load = 0.81

η_{gb} : Gearbox efficiency at full load = 0.85

η_{inv} : Inverter efficiency at full load = 0.96

For asynchronous generator and gearbox, we will consider 70% of full load losses as constant losses and other losses as load dependent losses. About the PM generator, the mechanical losses are less, we assumed the share of constant losses is 30% and the share of load dependent losses is 70%. In general, if   is a share constant loss factor, we have:

$$\eta = \frac{P_{out}}{P_{in}} = \frac{P_{out}}{P_{out}+P_{loss}} \tag{11}$$

$$P_{loss} = \frac{1-\eta}{\eta} P_{out} \tag{12}$$

$$P_{rot} = \alpha \frac{1-\eta}{\eta} P_{out} \tag{13}$$

Machine efficiency without constant losses is as follows:

$$\eta_v = \frac{P_{out}}{P_{out}+(1+\alpha)\frac{1-\eta}{\eta}(P_{out})} = \frac{\eta}{\eta+(1+\alpha)(1-\eta)} \tag{14}$$

The system can be analysed in three modes:

3.1. The First Case

Turbine is connected directly to a 10kW PM synchronous generator and the output is connected to a 50Hz grid with an inverter. Due to use of the inverter, wind turbine can operate at variable speed. The speed of the turbine can be in the range of 75rpm to 150rpm. From Equation (13) constant losses in this case are as follows:

$$P_{rot(PM1)} = 0.3 \times \frac{1-\eta_{PM}}{\eta_{PM}} \times 10kW = 0.26kW \tag{15}$$

The efficiency of the generator and the inverter without constant losses can be calculated from Equation (14) as follows:

$$\eta_{v(PM1)} = \frac{\eta_{PM}}{\eta_{PM}+(1-\alpha)(1-\eta_{PM})}\eta_{inv} = 0.905 \tag{16}$$

3.2. The Second Case

This case is similar with the first case with the exception that the synchronous generator power is 15 kW and it can be changed without any other characteristics changing.

$$\eta_{v(PM1)} = \eta_{v(PM2)} \tag{17}$$

$$P_{rot(PM2)} = 0.3 \times \frac{1-\eta_{PM}}{\eta_{PM}} \times 15kW = 0.39kW \tag{18}$$

3.3. The Third Case

Turbine is coupled to a 10 kW four-pole asynchronous generator through a gearbox with a ratio of 1:20 and three phase generator output is connected directly to the 50Hz grid. The speed of the turbine from non-load to full load varies between 750rpm to 78rpm. Constant losses are calculated as follows:

$$P_{rot(ind)} = 0.7 \times \frac{1-\eta_{ind}\eta_{gb}}{\eta_{ind}\eta_{gb}} \times 10kW = 3.167kW \tag{19}$$

$$n_{v(ind)} = \frac{\eta_{ind}\eta_{gb}}{\eta_{ind}\eta_{gb}+(1-\alpha)(1-\eta_{ind}\eta_{gb})} = 0.88 \tag{20}$$

$n_{v(ind)}$ shows the efficiency of the machine and gearbox without constant losses.

To calculate the annual energy, performance of the wind turbine and generator will be examined separately at low and high wind speeds:

Low wind speed: In this case, the generator works at a lower power than its nominal power and producing power is a function of wind speed and C_p. However, in the case where the PM generator is used, with suitable control of the inverter connected to the generator, C_p is a constant value equal to the maximum is 0.466.

For 10kW permanent magnet generator, the annual energy per kWh is calculated from Equation (9) and (10) at speeds between 3m/s and 8m/s as follows:

$$W_{PM1}|_{\frac{3m}{s}}^{\frac{8m}{s}} = \int_{t_1}^{t_1+\Delta t} P_{out}dt = \int_{t_1}^{t_1+\Delta t}(P_{mech} - P_{vLoss} - P_{rot(PM1)})dt = \int_{t_1}^{t_1+\Delta t}(P_{mech} - P_{vLoss})dt -$$
$$\int_{t_1}^{t_1+\Delta t} P_{rot(PM1)}dt = \eta_{v(PM1)}\int_{t_1}^{t_1+\Delta t} P_{mech}dt - P_{rot(PM1)}\Delta t|_{\frac{3m}{s}}^{\frac{8m}{s}} =$$
$$\frac{1}{2} \times \frac{8760}{1000} \times \eta_{v(PM1)}C_pA\rho \int_{\frac{3m}{s}}^{\frac{8m}{s}} v^3h(v)dv - 0.26 \times 8760 \times 0.4974 = 13888kWh \tag{21}$$

3m/s speed is selected based on (using Equation (10) and Figure 2), that turbine can produce power at least as P_{rot} and with increasing wind speed it can be injected power into the grid. In 8m/s speed producing power is equal to the nominal power of the generator and the turbine mechanical power output is equal to the output power with the total system losses.

$\Delta t|_{3m/s}^{8m/s}$ shows the duration of the wind speed between 3m/s to 8m/s and it can be calculated from Equation (9). This value is obtained at 4357 hours.

For 15kW permanent magnet generator, the annual energy per kWh is calculated at speeds between 3m/s and 9.5m/s as follows:

$$W_{PM2}|_{3m/s}^{9.5m/s} = \int_{t_1}^{t_1+\Delta t} P_{out}\, dt = \frac{1}{2} \times \frac{8760}{1000} \times \eta_{v(PM2)}C_pA\rho \int_{\frac{3m}{s}}^{\frac{9.5m}{s}} v^3h(v)dv - 0.39 \times 8760 \times$$
$$0.6 = 23954kWh \tag{22}$$

Because the power of this generator is higher wind speed is considered 9.5m/s. In this speed the mechanical power output of the turbine is equal to the nominal power of the generator and the system losses.

The energy calculation for asynchronous generator is a little different. The turbine speed is constant and with changing in the wind speed, TSR value is changing and according to

Figure 2 causes changing of C_p. Therefore C_p should be considered as a function of wind speed.

In 5.5m/s speed, turbine can produce power at least P_{rot} and gearbox losses and if the wind speed is a bit more to be able to be injected power into the grid. In 11m/s speed, mechanical power output of the turbine is equal to the nominal power of the generator and the total system losses.

If the wind speed is compared with the previous case, we can see in the previous case, the wind turbine has produced 17kW power with 9m/s speed, while in this case producing power is 14.5kW in 11m/s. The reason is that the permanent magnet generator using the inverter, possibility to change of the speed of the generator exists at maximum of C_p.

$$W_{ind}|_{5.5m/s}^{11m/s} = \int_{t_1}^{t_1+\Delta t} P_{out}dt = \eta_{v(ind)}\int_{t_1}^{t_1+\Delta t} P_{mech}dt - \int_{t_1}^{t_1+\Delta t} P_{rot(ind)}dt = \frac{1}{2} \times \frac{8760}{1000} \times$$

$$\eta_{v(ind)}A\rho\int_{\frac{5.5m}{s}}^{\frac{11m}{s}} C_p(\lambda)v^3h(v)dv - P_{rot(ind)}\Delta t|_{\frac{5.5m}{s}}^{\frac{11m}{s}} = 12764kWh \qquad (23)$$

High wind speed: In this case, with the controlled removal of the wind turbine from the wind direction (Furling), power from the wind is reduced and the output power of the generator is not exceeded from nominal value. In this case, the output power of the generator is equal to the nominal power. For 10kW permanent magnet generator we have:

$$W_{PM1}|_{8m/s}^{12m/s} = P_{out}\Delta t|_{\frac{8m}{s}}^{\frac{12m}{s}} = \eta_{inv} \times 10kW \times 8760 \times \int_{8m/s}^{12m/s} h(v)dv = 18112kWh \qquad (24)$$

Δt shows the duration of the wind speed for the mentioned speeds during the year.
For 15kW permanent magnet generator we have:

$$W_{PM2}|_{\frac{9.5m}{s}}^{\frac{12m}{s}} = P_{out}\Delta t|_{\frac{9.5m}{s}}^{\frac{12m}{s}} = \eta_{inv} \times 15kW \times 8760 \times \int_{\frac{9.5m}{s}}^{\frac{12m}{s}} h(v)dv$$

$$= 14145kWh \qquad (25)$$

For asynchronous generator we have:

$$W_{ind}|_{11m/s}^{12m/s} = P_{out}\Delta t|_{11m/s}^{12m/s} = 10kW \times 8760 \times \int_{11m/s}^{12m/s} h(v)dv = 3218kWh \qquad (26)$$

All of the above integrals are numerically calculated using MATLAB software.
The total annual energy delivered to the grid is as follows:
10kW PM generator:

$$W_{PM1} = W_{PM1}|_{3m/s}^{8m/s} + W_{PM1}|_{8m/s}^{12m/s} = 13888 + 18112 = 32000kWh \qquad (27)$$

15kW PM generator:

$$W_{PM2} = W_{PM2}|_{3m/s}^{9.5m/s} + W_{PM2}|_{9.5m/s}^{12m/s} = 23954 + 14145 = 38099kWh \qquad (28)$$

10kW asynchronous generator:

$$W_{PM1} = W_{ind}|_{5.5m/s}^{11m/s} + W_{ind}|_{11m/s}^{12m/s} = 12764 + 3218 = 15982kWh \qquad (29)$$

4. Compare the Cost and Value of Produced Energy Generators

Table 2 shows the costs and Table 3 shows the value of the energy produced in one year. Any one of the generators are not effective. The energy production of this type is economically for more than 45 penny per kilowatt-hour. In any case, in this paper, the advantages of each generator studied in comparison with other types.

Table 2. Approximate cost in various parts of the wind turbine

Estimated Costs		
Asynchronous ($)	Permanent Magnet ($)	
4800	4800	Blades
760	5200	10kW Generator
	8000	15kW Generator
1480	1480	Tower
600	600	Panel
--	4400	10kW Inverter
	6800	15kW Inverter
400	400	Measurement and control equipments
60	60	Cable
--	1000	Charging Control
520		Gearbox
800	800	Foundation
800	800	Installation costs
10220	19540	Total Cost 10kW
	24754	Total Cost 15kW

Table 3. Energy production cost in a year (13.34 p/kWh [10])

Type of generator	Energy production in a year (kWh)	Energy value in a year (penny)
PM 10kW	32000	426880
PM 15kW	38099	508241
Asynchronous	15982	213200

The cost of building of asynchronous is about 52% the cost of building of permanent magnet (10 kW). While the energy produced is about 50% of the permanent magnet. Although the difference is not too much.

From the continuity of energy production, asynchronous generator is started producing power from the 5.5m/s speed. In the case, permanent magnet generator can be produced power for 3m/s speed. This means that the asynchronous generator is stopped in most situations. Due to the asynchronous generator is operating at constant speed, all the changes of wind power appear as a tension in the blades and the structure that cause reduce the life of the structure the turbine. Therefore, in this respect, 10kW permanent magnet generator is a better choice.

About 15kW permanent magnet wind turbine, cost of building of this type is about 27% more than 10kW type. While produced energy is only 19% higher, which indicates the advantage of 10kW type.

However, it is important to note that these results are true only with a given Weibull curve in Figure 1. If the turbine installed in a region with the higher average wind speed, asynchronous generator would be more advantageous than 10kW permanent magnet type. However, in this case of 15kW can also be a suitable option.

5. Conclusion

Selecting the type of wind turbine should be done in each region. The results show that according to wind conditions, a 10kW permanent magnet generator is more advantageous in terms of energy production. Also this type of generator has the inherent advantages such as less tension in turbine and structures, greater coherence in injection power into the grid. It should be considered the energy advantage is not perennial and in a region with higher average wind speed, asynchronous generator is more advantageous due to lower initial cost.

References
[1] Ming Z, Sikaer A, Weiting G, Chen L. *Economic Analysis of the Stability in the Wind Turbine Selection.* Asia Pacific Power and Energy Engineering Conference. 2010: 1-4.
[2] Xing-jia Y, Zuo-xia X, Lei C, Hong-xia S. *Analysis of 1MW Variable Speed Wind Turbine Parameter Optimal Design Based on Cost Modeling Method.* IEEE Conference on Industrial Electronics and Applications. 2007; 15: 817-821.

[3] Yichun W, Ming D. *Optimal choice of wind turbine generator based on Monte-Carlo method*. International Conference on Electric Utility Deregulation and Restructuring and Power Technologies. 2008; 3: 2487-2491.

[4] Lu W, Ooi BT. *Multi-terminal LVDC system for optimal acquisition of power in wind-farm using induction generators*. Power Electronics Specialists IEEE Annual Conference. 2001; 32: 210-215.

[5] Wang F, Liu D, Zeng L. *Modeling and simulation of optimal wind turbine configurations in wind farms*. IEEE World Non-Grid-Connected Wind Power and Energy Conference. 2009: 1-5.

[6] Li H, Chen Z. *Optimal direct-drive permanent magnet wind generator systems for different rated wind speeds*. European Conference on Power Electronics and Applications. 2007: 1-10.

[7] Muljadi E, Butterfield CP. Pitch-Controlled Variable-Speed Wind Turbine Generation. *IEEE Transactions on Industry Applications*. 2001; 37: 240-246.

[8] Prasad RD, Bansal RC, Sauturaga M. Wind Energy Analysis for Vadravadra Site in Fiji Islands: A Case Study. *IEEE Transactions on Energy Conversion*. 2009; 24: 750-757.

[9] Fazelpour F, Soltani N, Rosen M. Wind resource assessment and wind power potential for the city of Ardabil, Iran. *International Journal of Energy and Environmental Engineering*. 2014: 1-8.

[10] http://www.wind-power-program.com/turbine_economics.htm. (accessed 27.03.15).

11

A New Approach to Adaptive Signal Processing

Muhammad Ali Raza Anjum
Department of Electrical Engineering, Army Public College of Management and Sciences,
Rawalpindi, Pakistan
e-mail: ali.raza.anjum@apcoms.edu.pk

Abstract

A unified linear algebraic approach to adaptive signal processing (ASP) is presented. Starting from just Ax=b, key ASP algorithms are derived in a simple, systematic, and integrated manner without requiring any background knowledge to the field. Algorithms covered are Steepest Descent, LMS, Normalized LMS, Kaczmarz, Affine Projection, RLS, Kalman filter, and MMSE/Least Square Wiener filters. By following this approach, readers will discover a synthesis; they will learn that one and only one equation is involved in all these algorithms. They will also learn that this one equation forms the basis of more advanced algorithms like reduced rank adaptive filters, extended Kalman filter, particle filters, multigrid methods, preconditioning methods, Krylov subspace methods and conjugate gradients. This will enable them to enter many sophisticated realms of modern research and development. Eventually, this one equation will not only become their passport to ASP but also to many highly specialized areas of computational science and engineering.

Keywords: *adaptive signal processing, adaptive filters, adaptive algorithms, Kalman filter, RLS, Wiener filter*

1. Introduction

In last few decades, Adaptive Signal Processing (ASP) has emerged into one of the most prolific areas in Electrical and Computer Engineering (ECE). It is based on the concept of intelligent systems that can automatically adapt to changes in their environment without the need for manual intervention [1]. ASP is gaining popularity with each passing day and is forming the basis of many key future technologies: including robots, gyroscopes, power systems, e-health systems, communication networks, audio and video technologies, etc [2-14]. As a result, ASP has become very important from both practical and pedagogical viewpoints. Therefore, its importance can hardly be over-emphasized.

But before one can enter into the realm of ASP, there are a lot of problems to face. To begin with, ASP lacks a unified framework. Numerous approaches to ASP can be found in literature and each of them tackles ASP in its own particular way. Secondly, the notation is mostly author specific. This makes it very difficult for the reader to understand even a single concept from two different manuscripts. Thirdly, almost every author leaves something out. For example, most authors drop the Kalman filter which has been recognized as one of most crucial topics in postgraduate research. Fourthly, authors make no attempt at making the concepts portable. Finally, the subject has a heavy probabilistic/statistical outlook which is overemphasized most of the time.

This work will provide a unified linear algebraic approach to ASP. Starting from just one equation in one unknown and one observation, all the key ASP algorithms - from Steepest Descent to Kalman filter - will be derived. During the derivation process, notation of algorithms will be kept uniform to make them consistent. Transitions from one algorithm to other will be kept systematic to make them portable. Probability and statistics shall not be invoked to make the algorithms accessible as well. The treatment of ASP will be entirely linear algebraic. Moreover, connection of ASP to other highly specialized domains like extended Kalman filter, reduced-rank adaptive filters, particle filters, multigrid methods, pre-conditioning methods, and Krylov subspace methods will also be established.

2. System Model

Let a system of linear equations be described as:

$$A_m x = b_m \tag{1}$$

With,

$$A_m = \begin{bmatrix} a^T_1 \\ a^T_2 \\ \vdots \\ a^T_m \end{bmatrix} \quad x = \begin{bmatrix} x_1 \\ x_2 \\ \vdots \\ x_n \end{bmatrix} \quad b_m = \begin{bmatrix} b_1 \\ b_2 \\ \vdots \\ b_m \end{bmatrix}$$

x represents the unknown system parameters. A_m contains the rows of the input data a_i's that are applied to this system. b_m comprises of the observations b_i's at the output of the system. x is an $(n \times 1)$ vector. A_m is an $(m \times n)$ matrix. b_m is an $(m \times 1)$ vector. According to this nomenclature, m inputs have been applied to the system with n unknown parameters and so far m outputs have been observed.

3. Least Mean Squares (LMS) Algorithm

Beginning with $m = 1$, there is one equation and one observation b_1.

$$a_1^T x = b_1 \tag{2}$$

Equation (2) seeks a vector x that has a projection of magnitude b_1 over the vector a_1. We begin with an arbitrary guess, say $x[k]$ at time-step k. Since our guess is purely arbitrary, the projection of $x[k]$ over a_1 may not be equal to b_1, i.e., $a_1^T x[k] \neq b_1$. There will be a difference. This difference can be denoted by an error $e[k]$.

$$e[k] = b_1 - a_1^T x[k] \tag{3}$$

$e[k]$ is the error in projection of $x[k]$ over a_1. It lies in the direction of a_1. The complete error vector $e[k]$ will be $e[k]a_1$. This error can be added to the initial guess $x[k]$ to rectify it. But the question is how much of it to add? Let there be a scalar μ in the range of $[0,1]$ such that the error vector is multiplied with μ before adding it to $x[k]$. A value of $\mu = 0$ indicates that no correction is required and the initial guess is accurate whereas a value of $\mu = 1$ indicates that a full correction is required.

$$x[k+1] = x[k] + \mu e[k]a_1 \tag{4}$$

$x[k+1]$ is the updated vector. Substituting Equation (3) in Equation (4).

$$x[k+1] = x[k] + \mu(b_1 - a_1^T x[k])a_1 \tag{5}$$

If k observations have been received up till the time-step,

$$x[k+1] = x[k] + \mu(b_k - a_k^T x[k])a_k \tag{6}$$

Equation (6) represents the famous LMS algorithm. The parameter μ in Eq. (6) is known as the step-size. It plays a crucial role in the convergence of the LMS algorithm. A very small step-size makes the algorithm crawl towards the true solution and hence terribly slows it down. This phenomenon is known as *lagging*. A large step-size makes the algorithm leap towards the true solution. Though it will make the algorithm much faster, movement in large steps never allows the algorithm to approach the true solution within a close margin. The algorithm keeps jumping around the true value but never converges to it and a further decrease in error becomes impossible. This phenomenon is known as *misadjustment* [1]. Effect of these phenomenons on the convergence properties of LMS algorithm are illustrated in Figure 1. These effects arise due to the fact that the step-size of LMS algorithm has to be adjusted manually. This is its major drawback. But inherent simplicity of LMS algorithm still keeps it much popular.

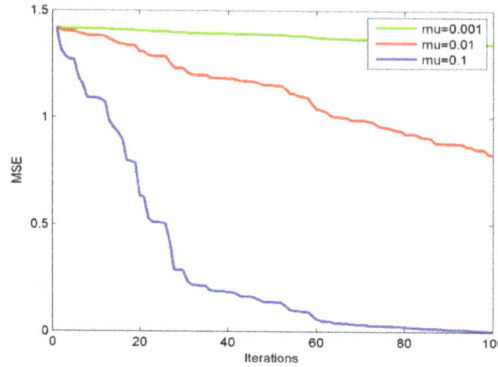

Figure 1. Comparison of the convergence properties of LMS algorithm with different step-sizes. LMS was employed to identify a system with an impulse response of length $n = 5$

4. Normalized Least Mean Squares (NLMS) Algorithm

NLMS provides an automatic adjustment in step-size. It is based on the criteria of selecting the best step-size for a given iteration. The term best is explained as follows. If error during the iteration is large, step-size is kept large so that the algorithm can quickly catch up with true solution. If the error decreases, step-size is lowered to allow the algorithm to zoom into the true solution. Hence, NLMS tries to select a step-size that minimizes the error in each iteration. In order to show how NLMS achieves it, we have to re-consider Equation (2). It represents the projection of vector x over the vector a_1 such that this projection has magnitude equal to b_1. But this is not the usual definition of dot product that represents orthogonal projections [15]. Actual definition requires the normalization of a_1 in Equation (2).

$$\frac{a_1^T x}{a_1^T a_1} = \frac{b_1}{a_1^T a_1} \tag{7}$$

Orthogonal projections make sure that error is orthogonal and hence, minimum. This in turn appears as a constraint on the step-size as we show now. Continuing in the similar fashion by choosing an arbitrary vector $x[k]$, the projection may not necessarily equal the right hand side and there will again be a difference.

$$\frac{b_1}{a_1^T a_1} - \frac{a_1^T x[k]}{a_1^T a_1} = \frac{1}{a_1^T a_1}(b_1 - a_1^T x[k]) = \frac{1}{a_1^T a_1}e[k]$$

Where,

$$e[k] = (b_1 - a_1^T x[k]) \tag{8}$$

The error vector will be,

$$e[k] = \frac{1}{a_1^T a_1}e[k]a_1 \tag{9}$$

Adding this correction to the original guess in order to improve it for the next iteration $x[k + 1]$,

$$x[k + 1] = x[k] + \frac{1}{a_1^T a_1}e[k]a_1 \tag{10}$$

Substituting Equation (8) in Equation (10),

$$x[k + 1] = x[k] + \frac{1}{a_1^T a_1}(b_1 - a_1^T x[k])a_1 \tag{11}$$

Or in general if k observations have been received up till the time-step,

$$x[k+1] = x[k] + \frac{1}{a_k^T a_k}(b_k - a_k^T x[k])a_k \qquad\qquad (12)$$

Equation (12) represents the NLMS algorithm. Comparing Equation (12) with Equation (6), it can be observed that,

$$\mu = \frac{1}{a_k^T a_k} \qquad\qquad (13)$$

In contrast to the LMS algorithm, step size of the NLMS algorithm is a variable which is automatically adjusted for each input row of data a_k according to its norm $a_k^T a_k$. By this automatic adjustment of step size, NLMS is able to avoid the problems of lagging and misadjustment that plague the LMS algorithm. Therefore, it has better convergence properties than LMS algorithm. Since NLMS achieves this advantage by the normalization step performed in Eq. (7), hence follows the name normalized LMS [1].

5. Kaczmarz Algorithm

Equation (11) for NLMS algorithm can be re-written as:

$$x[k+1] = x[k] + \frac{1}{a_k^T a_k}(b_k - a_k^T x[k])a_k \qquad\qquad (\text{Repeat})$$

This is the Kaczmarz equation. Kaczmarz algorithm was originally proposed for solving the under-determined systems $(m < n)$ [16]. Due to a limited number of rows, the Kaczmarz equation keeps jumping back to the first row after it has reached the last row until the solution converges. Hence, Kaczmarz algorithm is recurrent in terms of rows. Whereas in NLMS, new rows are added continuously due to the arrival of new data and the system, therefore, becomes over-determined, i.e., more rows than the columns. In this case, previous rows are never used. Therefore, NLMS algorithm is not recurrent. Otherwise both algorithms are identical.

6. Affine Projection (AP) Algorithm

Re-writing Equation (11) for NLMS algorithm,

$$x[k+1] = x[k] + a_1(a_1^T a_1)^{-1}(b_1 - a_1^T x[k]) \qquad\qquad (14)$$

NLMS algorithm tries to reduce the error in Equation (3) with respect to a single row of data a_1. The idea behind the AP algorithm is to choose a step-size that minimizes the error with respect to all the k rows of data that have been received up to time-step k to improve its convergence properties [17]. Hence, the vector a_1 in Equation (14) is replaced by a matrix A_k which contains all the k data rows.

$$x[k+1] = x[k] + A_k^T\left(A_k A_k^T\right)^{-1}(b_k - A_k x[k]) \qquad\qquad (15)$$

As long as the number of these data rows remains less than dimension of the system $(k < n)$, the system can be solved by Equation (15). Such a system is called an under-determined system [18]. It means that there are more unknowns than the number of equations which in turn implies that much less data is available. Equation (15) is known as AP algorithm and $A_k^T\left(A_k A_k^T\right)^{-1}$ is defined as the pseudoinverse for an under-determined system [18]. It is important to note that the pseudoinverse $A_k^T\left(A_k A_k^T\right)^{-1}$ is dependent on time-step k. Thus, the solution to Equation (15) is obtained by computing the term $A_k^T\left(A_k A_k^T\right)^{-1}$ at each time-step. In this way, AP algorithm may appear much more complex than NLMS and Kaczmarz algorithms. Also there is no indirect way of computing the pseudoinverse. However, these drawbacks are offset by its much faster convergence as compared to NLMS and Kaczmarz algorithms.

7. Recursive Least Squares (RLS) Algorithm

AP algorithm implies that the amount of data available is much less for the system to be full-determined $(k < n)$. However, this case seldom occurs in practice. On the contrary, more data keeps arriving with each time-step and the number of data rows exceeds the dimensions of the system $(k > n)$. As a result, the system becomes over-determined. There are more equations than the unknowns. In this case, the pseudoinverse for an under-determined systems $A_k^T(A_kA_k^T)^{-1}$ in Equation (15) is be replaced by the Least Squares pseudoinverse defined for an over-determined case $(A_k^TA_k)^{-1}A_k^T$ [18].

$$x[k + 1] = x[k] + (A_k^TA_k)^{-1}A_k^T(b_k - A_kx[k]) \qquad (16)$$

But before we can proceed with the iterative solution, we must consider $(A_k^TA_k)^{-1}$ term in Equation (16). This term will make Equation (16) converge in one iteration.

$$x[k + 1] = x[k] + (A_k^TA_k)^{-1}A_k^Tb_k - (A_k^TA_k)^{-1}A_k^TA_kx[k] = x[k] + x - x[k] \qquad (17)$$

Hence,

$$x[k + 1] = x \qquad (18)$$

But if we can compute this term beforehand, then the whole point of iterative solution becomes useless and the system can be directly solved in one step using Least Squares. Luckily, a lemma is available that can avoid the direct computation of $(A_k^TA_k)^{-1}$. This lemma is known as the *matrix inversion lemma* [18]. Here, we explain the great advantage achieved by this lemma over the AP algorithm where no such flexibility is available. We begin by examining the $A_k^TA_k$ term. This term can be decomposed as a sum of k rank-1 matrices.

$$A_k^TA_k = [a_1 \quad a_2 \quad \cdots \quad a_k]\begin{bmatrix} a^T_1 \\ a^T_2 \\ \vdots \\ a^T_k \end{bmatrix} = a_1a^T_1 + a_2a^T_2 + \cdots + a_{k-1}a^T_{k-1} + a_ka^T_k$$

$$A_k^TA_k = A_{k-1}^TA_{k-1} + a_ka^T_k \qquad (19)$$

The term $a_ka^T_k$ is known as the rank-1 *update-term* because it updates the $A_{k-1}^TA_{k-1}$ when a single new row of data a_k is added to the system at time-step k. The matrix inversion lemma computes the inverse $(A_k^TA_k)^{-1}$ by incorporating the rank-1 update $a_ka^T_k$ into the inverse $(A_{k-1}^TA_{k-1})^{-1}$. An identity matrix can be chosen as a starting candidate for $(A_{k-1}^TA_{k-1})^{-1}$ term.

$$(A_{k-1}^TA_{k-1} + a_ka_k^T)^{-1} = (A_{k-1}^TA_{k-1})^{-1} - \frac{(A_{k-1}^TA_{k-1})^{-1}a_ka_k^T(A_{k-1}^TA_{k-1})^{-1}}{\left(1+a_k^T(A_{k-1}^TA_{k-1})^{-1}a_k\right)} \qquad (20)$$

In this way, the system does not have to wait for all the rows of data to form its estimate x. Instead, the estimate is continually updated with the arrival of every new row of data. Also, the lemma seamlessly updates the estimate with every new observation and saves the toil of solving the system all over when new data arrives. Substituting Equation (19) in Equation (16) and modifying Equation (16) for one row of data at a time,

$$x[k + 1] = x[k] + (A_{k-1}^TA_{k-1} + a_ka^T_k)^{-1}a_k(b_k - a^T_kx[k]) \qquad (21)$$

Equation (21) is known as the RLS algorithm. Due to its dependency on matrix inversion lemma, RLS is much more complex than the rest of the algorithms discussed so far.

But despite that, it has been the most popular algorithm to date. Reasons for this are its ease of implementation, its excellent convergence properties, and its ability to update the estimate after every new observation without the need of solving the entire system of linear equations all over again. Figure 2 depicts the superior convergence properties of RLS algorithm as compared to the NLMS and LMS algorithms. Table 1 compares the computational complexity of the algorithms discussed so far [17].

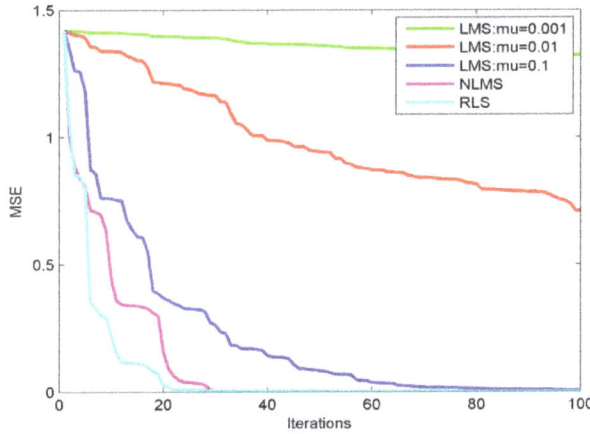

Figure 2. Comparison of the convergence properties of RLS, NLMS, and LMS algorithms

These algorithms were employed to identify a time-invariant system with an impulse response of length $n = 5$. System was over-determined

Table 1. Comparison of the computational complexity of various ASP algorithms

Algorithm	Complexity
LMS	$2N + 1$
NLMS	$3M + 2$
Kaczmarz	$3M + 2$
AF	N^2
RLS	N^2

8. Kalman Filter

All the algorithms that have been discussed up till now have a major underlying assumption. They provide an iterative solution to a time-invariant system. By time-invariant we mean that the vector x depicting the system parameters in Equation (1) remains unchanged during the iterative process. This assumption is relatively weak as it often happens in physical situations that the parameters of a system change during the convergence process, say for example in wireless communications [19]. Therefore, an algorithm must track the system during the transition process in addition to assuring the convergence within the intervals between the transitions. These are precisely the objectives of Kalman filter [20]. In order to show that how Kalman filter accomplishes it, we modify the system model in Equation (1) to incorporate the changes in nomenclature that arise due to the time varying nature of x.

$$A_{mn}x_n = b_m \tag{22}$$

A_{mn} is an $(m \times n)$ matrix. x is an $(n \times 1)$ vector. b_m is an $(m \times 1)$ vector. The system is over-determined $(m > n)$. m data rows and m observations have been received and the solution has been obtained using RLS algorithm in Equation (21). Let us assume that the system in Equation (22) changes its state from x_n to $\hat{x}_n$. This change of state will create a new set of unknowns $\hat{x}_n$ with the following linear relationship to previous unknowns x_n.

$$\widehat{x}_n = F_{nn}x_n + c_n \tag{23}$$

$\widehat{x}_n$ is an $(n \times 1)$ vector with n new system unknowns. F is an $(n \times n)$ matrix responsible for the change of state. It is also known as the *state-transition matrix*. c_n is an $(n \times 1)$ vector of constants. Re-writing Equation (23),

$$-F_{nn}x_n + \widehat{x}_n = c_n \tag{24}$$

Combining Equation (22) and (24) yields,

$$\begin{bmatrix} A_{mn} & 0_{nn} \\ -F_{nn} & I_{nn} \end{bmatrix} \begin{bmatrix} x_n \\ \widehat{x}_n \end{bmatrix} = \begin{bmatrix} b_m \\ c_n \end{bmatrix} \tag{25}$$

0_{nn} is an $(n \times n)$ matrix with all zero entries. I_{nn} is $(n \times n)$ identity matrix. Least Squares solution to Equation (25) is,

$$\begin{bmatrix} A^T_{mn} & -F^T_{nn} \\ 0_{nn} & I_{nn} \end{bmatrix} \begin{bmatrix} A_{mn} & 0_{nn} \\ -F_{nn} & I_{nn} \end{bmatrix} \begin{bmatrix} x_n \\ \widehat{x}_n \end{bmatrix} = \begin{bmatrix} A^T_{mn} & -F^T_{nn} \\ 0_{nn} & I_{nn} \end{bmatrix} \begin{bmatrix} b_m \\ c_n \end{bmatrix} \tag{26}$$

Adopting column-wise multiplication in Equation (26).

$$\left[\begin{bmatrix} A^T_{mn} \\ 0_{nn} \end{bmatrix} [A_{mn} \quad 0_{nn}] + \begin{bmatrix} -F^T_{nn} \\ I_{nn} \end{bmatrix} [-F_{nn} \quad I_{nn}] \right] \begin{bmatrix} x_n \\ \widehat{x}_n \end{bmatrix} = \begin{bmatrix} A^T_{mn} \\ 0_{nn} \end{bmatrix} b_m + \begin{bmatrix} -F^T_{nn} \\ I_{nn} \end{bmatrix} c_n \tag{27}$$

In order to find the value of new unknowns, we only need the inverse of the first term in the square brackets on the left hand side of Equation (27).

$$\begin{bmatrix} A^T_{mn} \\ 0_{nn} \end{bmatrix} [A_{mn} \quad 0_{nn}] + \begin{bmatrix} -F^T_{nn} \\ I_{nn} \end{bmatrix} [-F_{nn} \quad I_{nn}] \tag{28}$$

But we do not have to compute this inverse from right from start. From Equation (25), we observe that n new rows and n new columns have been added to the system in Equation (22) by F_{nn} and I_{nn} matrices. These extra rows and columns can be incorporated as rank-1 updates in the original matrix A_{mn}. This is because the zeros matrix 0_{nn} to the right of A_{mn} in Eq. (25) has no impact on the rows of A_{mn} other than increasing the length of individual rows from n to $2n$ by appendage n zeros at the end. This also leaves the product $A^T_{mn}A_{mn}$ unchanged. Only its dimensions have increased from $n \times n$ to $2n \times 2n$ with all the entries zeros expect the first $n \times n$ ones which are equal to $A^T_{mn}A_{mn}$.

$$\begin{bmatrix} A^T_{mn} \\ 0_{nn} \end{bmatrix} [A_{mn} \quad 0_{nn}] = \begin{bmatrix} A^T_{mn}A_{mn} & 0_{nn} \\ 0_{nn} & 0_{nn} \end{bmatrix} \tag{29}$$

Therefore, only the change in nomenclature is required to be precise. Otherwise the rest stays the same with this term. Therefore, we term the vector $[A_{mn} \quad 0]$ to $A_{m(2n)}$, meaning that $A_{m(2n)}$ has m rows each of length $2n$ with last n entries all zero. Similarly, we term $[-F_{nn} \quad I_{nn}]$ as $\widehat{F}_{n(2n)}$ because each of its rows is subjoined by the corresponding row of the identity matrix I_{nn}. Both the new and old unknowns in Equation (25) are jointly represented by x_{2n}. Equation (25) becomes,

$$\begin{bmatrix} A_{m(2n)} \\ \widehat{F}_{n(2n)} \end{bmatrix} x_{2n} = \begin{bmatrix} b_m \\ c_n \end{bmatrix} \tag{30}$$

Subsequently, Equation (28) becomes:

$$\begin{bmatrix} A^T_{mn} \\ 0_{nn} \end{bmatrix} [A_{mn} \quad 0_{nn}] + \begin{bmatrix} -F^T_{nn} \\ I_{nn} \end{bmatrix} [-F_{nn} \quad I_{nn}] = A^T_{m(2n)}A_{m(2n)} + \widehat{F}^T_{n(2n)}\widehat{F}_{n(2n)} \tag{31}$$

Matrix $\widehat{F}^T_{n(2n)} \widehat{F}_{n(2n)}$ in Equation (31) can be decomposed into a series of rank-one updates:

$$A^T_{m(2n)} A_{m(2n)} + \widehat{F}^T_{n(2n)} \widehat{F}_{n(2n)} = A^T_{m(2n)} A_{m(2n)} + \sum_{i=0}^{n} \widehat{f}_{i(2n)} \widehat{f}_{i(2n)}^T \qquad (32)$$

So the inverse required in Equation (26) can be written as:

$$\left[\begin{bmatrix} A^T_{mn} \\ 0_{nn} \end{bmatrix} [A_{mn} \quad 0_{nn}] + \begin{bmatrix} -F^T_{nn} \\ I_{nn} \end{bmatrix} [-F_{nn} \quad I_{nn}] \right]^{-1} = \left(A^T_{m(2n)} A_{m(2n)} + \sum_{i=0}^{n} \widehat{f}_{i(2n)} \widehat{f}_{i(2n)}^T \right)^{-1} \qquad (33)$$

The rank-1 updates of $\widehat{F}_{n(2n)}$ matrix can be used to form the inverse in Equation (33) using matrix inversion lemma. This inverse can then be used to from the pseudoinverse to compute the value of x_{2n}.

$$x_{2n} = \left(A^T_{m(2n)} A_{m(2n)} + \widehat{F}^T_{n(2n)} \widehat{F}_{n(2n)} \right)^{-1} \left(A^T_{m(2n)} b_m + \widehat{F}^T_{n(2n)} c_n \right) \qquad (34)$$

In this way, the values of the new unknowns can be determined without the need of explicitly computing the pseudoinverse. It is important to note that despite the system has changed state, new data measurements have not yet arrived. By solving for x_{2n} in advance, values of new unknowns can be *predicted* and the values of previous unknowns can be *smoothed*. Hence, Equation (34) is known as *prediction equation*. When the new measurement arrives, it is added as a new row of data to Equation (30).

$$\begin{bmatrix} A_{m(2n)} \\ \widehat{F}_{n(2n)} \\ a^T_{k(2n)} \end{bmatrix} x_{2n} = \begin{bmatrix} b_m \\ c_n \\ b_k \end{bmatrix} \qquad (35)$$

RLS can now be invoked to update the predicted vector $x_{2n}[k]$ in the light of new measurements.

$$x_{2n}[k+1] = $$
$$x_{2n}[k] + \left(A^T_{m(2n)} A_{m(2n)} + \widehat{F}^T_{n(2n)} \widehat{F}_{n(2n)} + a_{k(2n)} a^T_{k(2n)} \right)^{-1} a_{k(2n)} \left(b_k - a^T_{k(2n)} x_{2n}[k] \right) \qquad (36)$$

Again the matrix inversion lemma can be used to compute the term $\left(A^T_{m(2n)} A_{m(2n)} + FTn(2n)Fn(2n)+ak(2n)aTk(2n)\right)-1$ by employing the rank-1 update $ak(2n)aTk(2n)$ in the term $A^T_{m(2n)} A_{m(2n)} + \widehat{F}^T_{n(2n)} \widehat{F}_{n(2n)}$ calculated in Equation (32). Equation (36) is known as *update equation*. Following term in update equation,

$$\left(A^T_{m(2n)} A_{m(2n)} + \widehat{F}^T_{n(2n)} \widehat{F}_{n(2n)} + a_{k(2n)} a^T_{k(2n)} \right)^{-1} a_{k(2n)} \qquad (37)$$

Is known as *Kalman gain*. As more and more measurements arrive, Equation (36) is re-run to update the predicated estimate. In case the state of the system changes again, new unknowns can be added to the system in the same manner as described before in Equation (25) and the same procedure can be repeated. Though we have kept the information of old states together with the new states by the process of *smoothing*, it is not always necessary to do so. As soon as the measurements arrive and the estimated is update, old estimates can be thrown away altogether to keep the information about the most recent state only. This change of states which arises due to the time-varying nature of the system is the only feature that distinguishes Kalman filter from RLS. Otherwise, Kalman filter is similar to RLS because they perform identically for a time-invariant system. For this reason, Kalman filter is often called the time varying RLS [21].

9. Steepest Descent (SD) Algorithm
Equation (5) for LMS algorithm can be expanded as:

$$x[k+1] = x[k] + \mu(a_k b_k - a_k a_k{}^T x[k]) \tag{38}$$

Instead of taking a row at a time in Equation (38), SD is interested in the average of all the k rows that have been received so far.

$$\frac{1}{k}(a_1 a^T{}_1 + a_2 a^T{}_2 + \cdots + a_k a^T{}_k) = \frac{1}{k} a_1 a^T{}_1 + \frac{1}{k} a_2 a^T{}_2 + \cdots + \frac{1}{k} a_k a^T{}_k \tag{39}$$

$\frac{1}{k}$ represents the probability of each row as each row is equally likely.

$$P(a_1)a_1 a^T{}_1 + P(a_2)a_2 a^T{}_2 + \cdots + P(a_k)a_k a^T{}_k = \sum_{i=1}^{k} P(a_i)a_i a^T{}_i = E\{a_i a^T{}_i\} \tag{40}$$

E is an expectation operator and stands for expected or average value. Similarly for $a_k b_k$ term in Equation (39),

$$\frac{1}{k}a_1 b_1 + \frac{1}{k}a_2 b_2 + \cdots + \frac{1}{k}a_k b_k = P(a_1)a_1 b_1 + P(a_2)a_2 b_2 + \cdots + P(a_k)a_k b_k \tag{41}$$

$$E\{a_i b_i\} = \sum_{i=1}^{k} P(a_i)a_i b_i \tag{42}$$

Substituting Equation (42) and (40) in Equation (38),

$$x[k+1] = x[k] + \mu(E\{a_i b_i\} - E\{a_i a^T{}_i\}x[k]) \tag{43}$$

Whereas,

$$R = E\{a_i a^T{}_i\} \tag{44}$$

$$P = E\{a_i b_i\} \tag{45}$$

R is known as the *auto-correlation matrix* and P as the *cross-correlation matrix* [1]. Substituting Equation (44) and (45) in Equation (43),

$$x[k+1] = x[k] + \mu(P - Rx[k]) \tag{46}$$

Equation (46) represents the Steepest Descent algorithm. This equation differs from Equation (6) of LMS algorithm only in terms of R and P matrices. They are formed in SD algorithm by taking the average of all the available data rows as depicted in Equation (40) and (42). LMS algorithm, on the other hand, only considers one row of data at a time. It has been demonstrated that if the LMS algorithm is run repeatedly for a given problem, then on the average its performance will be equal to that of LMS algorithm [1]. But R and P matrices are usually not available beforehand and computing them at run-time can be expensive. Therefore, LMS algorithm is much more popular than SD algorithm due to its simplicity.

10. MMSE and Least Squares Wiener Filters
As Equation (46) represents the recursion equation for SD algorithm, $(P - Rx[k])$ must be the gradient.

$$\nabla = P - Rx[k] \tag{47}$$

Setting the gradient in Equation (47) to zero and solving directly for x,

$$Rx = P \tag{48}$$

Equation (48) is known as Minimum Mean Square Error (MMSE) Wiener-Hop equation. x is now called MMSE Wiener Filter [1]. Expanding Equation (46),

$$E\{a_i a^T_i\}x = E\{a_i b_i\} \tag{49}$$

As can be observed from Equation (39) and (40),

$$E\{a_i a^T_i\} = \sum_{i=1}^{k} P(a_i) a_i a^T_i = \frac{1}{k}(a_1 a^T_1 + a_2 a^T_2 + \cdots + a_k a^T_k) = \frac{1}{k} A_k^T A_k \tag{50}$$

Similarly from Equation (41) and (42),

$$E\{a_i b_i\} = \frac{1}{k}(a_1 b_1 + a_2 b_2 + \cdots + a_k b_k) = \frac{1}{k}\begin{bmatrix} a_1 & a_2 & \cdots & a_k \end{bmatrix}\begin{bmatrix} b_1 \\ b_2 \\ \vdots \\ b_k \end{bmatrix} = \frac{1}{k} A_k^T b_k \tag{51}$$

Substituting Equation (50) and (51) in Equation (48),

$$A_k^T A_k x = A_k^T b_k \tag{52}$$

Equation (52) is known as the Least Square Wiener-Hop equation. x is now called Least Squares Wiener Filter [1].

11. Relationship with other Areas
This section establishes the relationship of the work presented in this article to highly specialized fields that are currently the focus of research and development in ECE and applied mathematics. This will enable the readers to explore further opportunities for research and innovation.

11.1. Reduced Rank Adaptive Filters
We begin our discussion of reduced-rank adaptive filters with Equation (52).

$$x = \left(A_k^T A_k\right)^{-1} A_k^T b_k \tag{53}$$

Least Squares Wiener filter x in Equation (52) requires the inversion of $A_k^T A_k$ matrix. The matrix $A_k^T A_k$ is also known as *covariance matrix*. For it to be invertible, it must have full-rank. But this assumption of full-rank has been strongly questioned by the proponents of reduced rank adaptive filters [22]. They claim that such full-rank is not available often in practice, say for example in sensor array processing. This claim not only puts the invertibility of $A_k^T A_k$ matrix in question but also the challenges the entire possibility of finding the LS wiener filter. However, the rank of $A_k^T A_k$ can be easily analyzed because it is a symmetric matrix. A symmetric matrix can be decomposed into a set of real eigenvalues and orthogonal eigenvectors [15].

$$A_k^T A_k = Q \Lambda Q^T \tag{54}$$

Λ is an $(n \times n)$ diagonal matrix of eigenvalues. Q is an $(n \times n)$ matrix of eigenvectors. For a matrix to be of full-rank, none of its eigenvalues should be zero. Even if one of them is zero, the matrix becomes non-invertible. In this case, a Moore and Penrose's pseudoinverse can be formed using Equation (54) [18]. In order to explain the key idea here, we have to concentrate our attention on Q matrix for a moment. Q is an orthogonal matrix, i.e., $Q^T Q = I$. This implies that inverse of Q is equal to its transpose, i.e., $Q^{-1} = Q^T$. Inverting $A_k^T A_k$ would mean,

$$\left(A_k^T A_k\right)^{-1} = (Q \Lambda Q^T)^{-1} = (Q^T)^{-1} \Lambda^{-1} Q^{-1} = Q \Lambda^{-1} Q^T \tag{55}$$

Hence, to form the inverse $\left(A_k^T A_k\right)^{-1}$ we only need to invert the eigenvalues. But since the matrix $A_k^T A_k$ is rank-deficient, at least one of its eigenvalues must be zero. Inverting such an eigenvalue will create problems. Idea behind the Moore and Penrose's pseudoinverse is to leave such an eigenvalue un-inverted while inverting the rest [18].

$$\left(A_k^T A_k\right)^+ = Q \Lambda^{-1} Q^T \qquad (56)$$

$\left(A_k^T A_k\right)^+$ indicates the Moore and Penrose's pseudoinverse. A reduced rank LS Wiener filter can then be obtained by:

$$x = \left(A_k^T A_k\right)^+ A_k^T b_k \qquad (57)$$

Similarly, the argument can be extended to RLS algorithm which also requires the inversion of the covariance matrix.

$$x[k+1] = x[k] + \left(A_{k-1}^T A_{k-1} + a_k a^T_k\right)^{-1} a_k (b_k - a^T_k x[k]) \qquad (\text{Repeat})$$

As we know from Equation (19),

$$A_k^T A_k = A_{k-1}^T A_{k-1} + a_k a^T_k \qquad (\text{Repeat})$$

Therefore,

$$x[k+1] = x[k] + \left(A_k^T A_k\right)^{-1} a_k (b_k - a^T_k x[k]) \qquad (58)$$

Here again, if the rank of covariance matrix is not full, pseudo inverse $\left(A_k^T A_k\right)^+$ can be used in place of $\left(A_k^T A_k\right)^{-1}$.

$$x[k+1] = x[k] + \left(A_k^T A_k\right)^+ a_k (b_k - a^T_k x[k]) \qquad (59)$$

However, this is not the only case argued by the theory of reduced rank adaptive filters. Now that we can form the pseudoinverse, rank of the covariance matrix can be deliberately reduced if some of its eigenvalues are very small. In this way, larger eigenvalues of the covariance matrix can be given preference which have a larger impact on the system performance. This idea is known by the name of compressing [22]. It allows the removal of unnecessary data which has no or very less impact on systems performance. As a result, data processing ability of system is greatly improved.

11.2.   Extended Kalman Filter for Non-Linear Control/Signal Processing
In non-linear systems, the system model in Eq. (22) and state transition model in Equation (23) become non-linear. In this case, the system matrix A_{mn} and the state transition matrix F_{nn} for Kalman filter are obtained from a first order Taylor series expansion of these models, a technique called *linearization*. Rest stays the same and Equation (36) is now called the extended Kalman filter (EKF) [23]. It is the most popular filter for tackling non-linear filtering problems and forms the core of non-linear signal processing, non-linear control theory, and robotics.

11.3.   Particle Filters and Bayesian Signal Processing
If more terms of the Taylor series are retained during the linearization process, higher order EKF is obtained which performs much better than its predecessor. But these extra terms introduce additional complexity. Hence, there is a classic tradeoff between complexity and accuracy. An alternative approach to this problem is to use Monte Carlo techniques to estimate system parameters for better performance. In this case, system and state transition matrices in Equation (22) and (23) are replaced by probability estimates and Equation (36) is now termed as a particle filter. It is major driving concept behind particle filtering theory and Bayesian signal processing [24].

11.4. Multigrid Methods for Boundary Value Problems
Repeating Equation (1) for the system model,

$$A_m x = b_m \qquad \text{(Repeat)}$$

If we make an arbitrary guess $x[k]$ for the unknown x, then the right hand side may not be equal to the left hand side and there will be a difference. This ensuing difference is known as the residue term r_m in multigrid methods.

$$r_m = b_m - A_m x[k] \qquad (60)$$

Multiplying both sides of Equation (60) with the psuedoinverse $(A^T_m A_m)^{-1} A^T_m$,

$$(A^T_m A_m)^{-1} A^T_m r_m = (A^T_m A_m)^{-1} A^T_m b_m - (A^T_m A_m)^{-1} A^T_m A_m x[k] \qquad (61)$$

From Equation (61), it follows,

$$(A^T_m A_m)^{-1} A^T_m r_m = x - x[k] = e \qquad (62)$$

Or,

$$x = x[k] + e \qquad (63)$$

e is the error between the true solution and our guess. It is termed as error in the estimate or simply the error. Re-arranging Equation (62),

$$A_m e = r_m \qquad (64)$$

Equation (64) represents the relationship between the error and the residue. It is identical in structure to the system model in Equation (1). Residue r_m takes place of the observations b_m and the error e becomes the unknown. However trivial it may seem, it signifies a very important concept behind multigrid methods. An initial guess $x[k]$ for the estimate x is made. This guess is used to obtain the residue r_m in Equation (60). The residue can then be used to solve for the error e in Equation (64). Once Equation (64) is solved, e can be added back to $x[k]$ to find the true estimate x. Now we move on to explain the second driving concept behind the multigrid methods. Equation (6) for LMS algorithm can be used to solve Equation (1).

$$x[k+1] = x[k] + \mu(b_k - a_k^T x[k]) a_k \qquad \text{(Repeat)}$$

It can also be used to solve Equation (62).

$$e[k+1] = e[k] + \mu(r_k - a_k^T e[k]) a_k \qquad (65)$$

This is because Equation (64) is similar in structure to Equation (1). Initially Eq. (6) is run to obtain $x[k]$. Then $x[k]$ is plugged into Eq. (60) to obtain r_m. After that Equation (65) is run to directly improve the error in the estimate e. It happens in practice that error constitutes a mixture of high and low frequency oscillations, say for example in the numerical solution of Laplace Equation [18]. High frequency oscillations in the error die out quickly as manifested in fast initial convergence of the algorithm. Low frequency oscillations, on the other hand, tend to linger on. These low frequency errors are responsible for settling the algorithm in the steady state in which even after significant number of iterations, reduction in error is negligible. It is well-known from multirate signal processing that down-sampling a signal increases its frequency [25]. So by down-sampling the error, these low frequency oscillations can be converted to high frequency ones. Afterwards, the algorithm can be re-run to damp them out. When the error ceases to decrease, it can be further down-sampled and the algorithm is once again re-run. In this way, error can be reduced to any desirable tolerance in multiple cycles and at multiple scales. These multiple scales are known as multigrids and, hence, follows the name multigrid methods. Multigrid methods are the power house of boundary value problems arising in computational electromagnetics, computational fluid dynamics, statistical physics, wavelet theory, applied mathematics, and many other branches of computational science and engineering [18].

11.5. Preconditioning Methods for Large Linear Systems
Consider Equation (6) for LMS algorithm once again,

$$x[k+1] = x[k] + \mu a_k (b_k - a_k^T x[k])$$ (Repeat)

The term $(b_k - a_k^T x[k])$ is the residue for k-th row of A_k matrix in Equation (1). The residue is generated as soon as the k-th data row becomes available. This is ideal for run-time or on the fly operation. But if all the input and output data is available beforehand, say for example pre-recorded in a laboratory, then the entire A_m matrix and the b_m vector can be used at once in Equation (6) instead of the individual data rows and observations.

$$x[k+1] = x[k] + \mu A^T_m (b_m - A_m x[k])$$ (66)

We have used A_m and b_m instead of A_k and b_k because only m input data rows and m output observations are available in the lab. So, we have to find x that agrees with all the available observations. No new measurements will be available. Now the question arises about the choice of best step-size for Equation (66). If we want to solve Equation (66) in one-step, the ideal step-size would be $\mu = (A^T_m A_m)^{-1}$.

$$x[k+1] = x[k] + (A^T_m A_m)^{-1} A^T_m (b_m - A_m x[k])$$ (67)

Expanding Equation (67),

$$x[k+1] = x[k] + (A^T_m A_m)^{-1} A^T_m b_m - (A^T_m A_m)^{-1} A^T_m A_m x[k]$$ (68)

Finally,

$$x[k+1] = x[k] + x - x[k] = x$$ (69)

So the system will converge in one step. Observe that for this to happen we practically need to multiply the residue term $(b_m - A_m x[k])$ in Equation (67) with the pseudoinverse $(A^T_m A_m)^{-1} A^T_m$. But if the pseudoinverse $(A^T_m A_m)^{-1} A^T_m$ is available beforehand, whole point of iterative solution becomes moot. Instead the system can be solved directly. Here the preconditioning methods come to rescue. These methods suggest that instead of multiplying the pseudoinverse $(A^T_m A_m)^{-1} A^T_m$ with the residue term, we can choose another matrix P that is much easier to invert.

$$x[n+1] = x[n] + P^{-1}(b_m - A_m x[n])$$ (70)

P can be the diagonal part of the pseudoinverse or may constitute its lower-triangular portions. Former choice is known as the *Jacobi iteration* and the latter one as the *Gauss-Seidel iteration*. *Successive Over Relaxation (SOR)* is a combination of both [18]. Anyhow, the idea is to find a P that is much easier to invert. Solving Equation (70),

$$Px[n+1] = Px[n] + (b_m - A_m x[n])$$ (71)

Or,

$$Px[n+1] = (P - A_m)x[n] + b_m$$ (72)

Equation (72) is known as the *preconditioning equation*. P is known as the *preconditioning matrix*. The process $(P - A)$ is termed as *splitting*. Preconditioning methods are popular for solving large linear systems that are too expensive for traditional methods based on matrix factorizations.

11.6.   Krylov Subspace Methods and Conjugate Gradients for Optimization and Control
If identity matrix is chosen as the precondition matrix P and b_m as the initial guess $x[n]$, Equation (72) becomes,

$$x[n+1] = (I - A_m)b_m + b_m = 2b_m - A_m b_m \qquad (73)$$

Iterating Equation (73) for the second time,

$$x[n+2] = (I - A_m)x[n+1] + b_m = 3b_m - 3A_m b_m + A^2{}_m b_m \qquad (74)$$

$b_m, A_m b_m, ..., A^n{}_m b_m$ constitute the basis vectors for the Krylov subspace. According to Equation (74), x can be obtained by a linear combination of Krylov basis vectors. Hence, the solution to Equation (1) can be found in Krylov subspace. This is the driving concept behind Krylov subspace methods. Advanced Krylov subspace methods include conjugate gradient method and minimum Residual methods: MINRES and GMRES [18]. These methods are central to optimization theory and automatic control systems.

12. Discussion
Various authors have tried to address the problems mentioned in the Introduction section, though in part only. Widrow & Stearn [26], for example, have tried to present a broad conceptual introduction to the basic algorithms. Their treatment of subject is light on mathematics and is definitely not for someone who wishes to implement these algorithms. Boroujeny [1] has attempted to make these algorithms more implementable by providing a concrete mathematical approach. However, the usual clarity and the knack of exposition of the author just disappear as the algorithms become more advanced. For example, author just restates the major results in RLS filtering in terms of his own notation and drops the Kalman filter entirely. Poularikas [27] has made an effort to make the basic algorithms accessible by using the simulation approach. Yet again, Kalman filter is entirely left out and RLS is barely touched. Diniz [28] has provided a detailed treatment of these algorithms by following a statistical/linear algebraic approach to the subject. But when it comes to Kalman filter, the author switches to state space paradigm of control theory for a brief introduction to the topic. Strang [20, 21] has tried to tackle the RLS and Kalman filtering problem and has endeavored to bring it to linear algebraic perspective. But the author entirely leaves out the Wiener filters, Kaczmarz, LMS, and NLMS algorithms. Dwight [29] has tried to categorize the basic algorithms in two different paradigms, the least squares paradigm and MMSE paradigm using linear algebraic approach. But when it comes to the exposition of Kalman filter, the author completely switches to Wiener's system theory and statistical signal processing paradigm resulting in almost no connection between his earlier and later results. To unify adaptive signal processing and statistical signal processing platforms, an effort has been made by Manolakis et el [30]. But the book has rather obscure notation and the topics are almost inaccessible without familiarization with it. Chong [16] provides a treatment of only RLS and Kaczmarz algorithms from optimization theoretic viewpoint. Whereas Hayes [31], Syed [32], Haykin [33], and Candy [24] make no attempt at all at making these algorithms accessible.

As we have been through this journey, we have discovered two things that have shaped our methodology. Firstly, no matter how hidden it is behind all the veils just mentioned, it is all linear algebra. Secondly, from the basic Steepest Descent algorithm to fairly advanced Kalman filter, one and only one equation is involved. This equation assumes a different name and wears a different mantle of symbols in each algorithm. If somehow this link can be discovered, a relationship can be established among all the algorithms. Then the transition from one algorithm to the other will be more logical, more systematic, and much easier to understand. This will provide an opportunity to focus more on the underlying problem, the reason for the progression from one algorithm to the other, and the advantage provided by the new algorithm over the previous one. Also their notation can be made more uniform which will generate certain terseness in their explanation. In this way, one approach can be picked and followed to the end without the need for switching between different subjects. We have adopted a linear algebraic approach for this purpose. It is the simplest possible approach in a sense that it asks for no technical introduction, making it possible for everyone to benefit. We have started from the simplest possible case, one equation in one unknown and one observation. From this one equation, we have derived all the key algorithms from SD to Kalman filter while leaving nothing out. Notation was kept consistent throughout. Transitions from one algorithm to the other were logical. They were also systematic to make the concepts portable. Calculus, Probability, and Statistics were not invoked in order to make them accessible as well. The treatment was entirely linear algebraic. Each step was explained but the explanations were kept concise in order to prevent the manuscript from becoming too voluminous.

13. Conclusion

Our work provides a uniform approach to students, researchers, and practitioners from different academic backgrounds working in different areas. Anyone can learn these key algorithms without any previous knowledge. The understanding of readers will be enhanced and their efforts will be minimized as they will see just one equation in action. It will enable them to focus more on the concept rather than the symbols, helping them to master the essentials in minimum amount of time. As there is only one equation to deal with, it will also ease their burden of programming. They can program it without the need of special programming skills. They will be able to appreciate how just one equation has been the centre of research and development for last many decades and how it has vexed scientists and engineers in many different fields. In fact, to demonstrate the benefit of this synthesis, an entire area in applied mathematics known as multigirid methods that aim to solve differential equations by operating at multiple resolutions, stems from this one equation. Other specialized domains like non-linear control and particle filter theory also take the lead from this one equation. Finally, a better understanding of these algorithms will ultimately lead to a better understanding of image processing, video processing, wireless communications, pattern recognition, machine learning, optimization theory and many other subjects that utilize these core concepts. For anyone, this one equation will eventually be their passport to the realm of computational science and engineering. For students, this one equation will save them the toil of going through these algorithms again and again in various courses where they just change their name and switch their clothes. For faculty, this one equation will spare their efforts to teach these algorithms repeatedly in various courses. For universities, this one equation will enable them to develop a consolidated course based on these algorithms that are central to so many subjects in the ECE and applied mathematics.

References
[1]   B Farhang-Boroujeny. Adaptive Filters: Theory and Applications. John Wiley & Sons. 2013.
[2]   P Avirajamanjula, P Palanivel. Corroboration of Normalized Least Mean Square Based Adaptive Selective Current Harmonic Elimination in Voltage Source Inverter using DSP Processor. *International Journal of Power Electronics and Drive Systems (IJPEDS)*. 2015; 6.
[3]   Y Zheng-Hua, R Zi-Hui, Z Xian-Hua, L Shi-Chun. Path Planning for Coalmine Rescue Robot Based on Hybrid Adaptive Artificial Fish Swarm Algorithm. *TELKOMNIKA Indonesian Journal of Electrical Engineering*. 2014; 12: 7223-7232.
[4]   F Wang, Z Zhang. An Adaptive Genetic Algorithm for Mesh Based NoC Application Mapping. *TELKOMNIKA Indonesian Journal of Electrical Engineering*. 2014; 12: 7869-7875.
[5]   X Yan, J Li, Z Li, Y Yang, C Zhai. Design of Adaptive Filter for Laser Gyro. *TELKOMNIKA Indonesian Journal of Electrical Engineering*. 2014; 12: 7816-7823.
[6]   J Mohammed. Low Complexity Adaptive Noise Canceller for Mobile Phones Based Remote Health Monitoring. *International Journal of Electrical and Computer Engineering (IJECE)*. 2014; 4: 422-432.
[7]   KF Akingbade, IA Alimi. Separation of Digital Audio Signals using Least Mean Square LMS Adaptive Algorithm. *International Journal of Electrical and Computer Engineering (IJECE)*. 2014; 4: 557-560.
[8]   J Mohammed, M Shafi. An Efficient Adaptive Noise Cancellation Scheme Using ALE and NLMS Filters. *International Journal of Electrical and Computer Engineering (IJECE)*. 2012; 2: 325-332.
[9]   J Mohammed. A Study on the Suitability of Genetic Algorithm for Adaptive Channel Equalization. *International Journal of Electrical and Computer Engineering (IJECE)*. 2012; 2: 285-292.

[10] J Huang, J Xie, H Li, G Tian, X Chen. Self Adaptive Decomposition Level Denoising Method Based on Wavelet Transform. *TELKOMNIKA Indonesian Journal of Electrical Engineering.* 2012; 10: 1015-1020.

[11] N Makwana, M SPIT, N Mishra, B Sagar. Hilbert Transform Based Adaptive ECG R Peak Detection Technique. *International Journal of Electrical and Computer Engineering (IJECE).* 2012; 2: 639-643.

[12] Z Chen, U Qiqihar, X Dai, L Jiang, C Yang, B Cai. Adaptive Iterated Square Root Cubature Kalman Filter and Its Application to SLAM of a Mobile Robot. *TELKOMNIKA Indonesian Journal of Electrical Engineering.* 2013; 11: 7213-7221.

[13] Z Ji, Adaptive Cancellation of Light Relative Intensity Noise for Fiber Optic Gyroscope. *TELKOMNIKA Indonesian Journal of Electrical Engineering.* 2013; 11: 7490-7499.

[14] Q Liu. An Adaptive Blind Watermarking Algorithm for Color Image. *TELKOMNIKA Indonesian Journal of Electrical Engineering.* 2013; 11: 302-309.

[15] G Strang. Introduction to Linear Algebra. Wellesley-Cambridge Press. 2003.

[16] EK Chong, SH Zak. An Introduction to Optimization. John Wiley & Sons. 2013; 76.

[17] KA Lee, WS Gan, SM Kuo. Subband Adaptive Filtering: Theory and Implementation. John Wiley & Sons. 2009.

[18] G Strang. Computational Science and Engineering. Wellesley-Cambridge Press. 2007.

[19] TS Rappaport. Wireless Communications: Principles and Practice. Prentice hall PTR New Jersey. 1996; 2.

[20] G Strang, K Borre. Linear Algebra, Geodesy, and GPS. Wellesley-Cambrdige Press. 1997.

[21] G Strang. Introduction to Applied Mathematics. Wellesley-Cambridge Press. 1986; 16.

[22] JS Goldstein, IS Reed. Reduced-rank Adaptive Filtering. *IEEE Transactions on Signal Processing.* 1997; 45: 492-496.

[23] J Mochnác, S Marchevský, P Kocan. *Bayesian Filtering Techniques: Kalman and Extended Kalman Filter Basics.* In Radioelektronika, 2009. RADIOELEKTRONIKA'09. 19th International Conference. 2009: 119-122.

[24] JV Candy. Bayesian Signal Processing: Classical, Modern and Particle Filtering Methods. John Wiley & Sons. 2011; 54.

[25] G Strang, T Nguyen. Wavelets and Filter Banks. Wellesley-Cambrdige Press. 1996.

[26] B Widrow, SD Stearns. Adaptive Signal Processing. Prentice-Hall, Inc. 1985; 1.

[27] AD Poularikas, ZM Ramadan. Adaptive Filtering Primer with MATLAB. CRC Press. 2006.

[28] PS Diniz. Adaptive Filtering. Springer. 1997.

[29] DF Mix. Random Signal Processing. Prentice-Hall, Inc. 1995.

[30] DG Manolakis, VK Ingle, SM Kogon. Statistical and Adaptive Signal Processing: Spectral Estimation, Signal Modeling, Adaptive Filtering, and Array Processing. Artech House Norwood. 2005; 46.

[31] MH Hayes. Statistical Digital Signal Processing and Modeling. John Wiley & Sons. 1996.

[32] AH Sayed. Fundamentals of Adaptive Filtering. John Wiley & Sons. 2003.

[33] SS Haykin. Adaptive Filter Theory. Pearson Education India. 2008.

12

Influencing Power Flow and Transient Stability by Static Synchronous Series Compensator

Md. Imran Azim, Md. Abdul Wahed, Md. Ahsanul Haque Chowdhury
Departement of Electrical and Electronic Engineering,
Rajshahi University of Engineering and Technology (RUET)
e-mail: imran.azim89@gmail.com

Abstract

In the present world, modern power system networks, being a complicated combination of generators, transmission lines, transformers, circuit breakers and other devices, are more vulnerable to various types of faults causing stability problems. Among these faults, transient fault is believed to be a major disturbance as it causes large damage to a sound system within a certain period of time. Therefore, the protection against transient faults, better known as transient stability control is one of the major considerations for the power system engineers. This paper presents the control approach in the transmission line during transient faults by means of Static Synchronous Series Compensator (SSSC) in order to stabilize Single Machine Infinite Bus (SMIB) system. In this paper, SSSC is represented by variable voltage injection associated with the transformer leakage reactance and the voltage source. The comparative results depict that the swing curve of a system increases monotonically after the occurrence of transient faults. However, SSSC is effective enough to make it stable after a while.

Keywords: SMIB, power angle curve, transient stability, SSSC, swing curve

1. Introduction

With the passage of time, the demand of electricity, being considered as a vital source of energy, has been leapt due to the fact that the number of consumers is growing day by day. This situation urges the necessity of adding more transformers, transmission lines, switchgear equipment and distribution lines indicating a complex power system. It is needless to say the possibility of his power system network to be affected by the faults is very high. As a result, stability problem is a key concern as far as modern power system network goes [1-3].

Stability refers to the tendency of a power system to develop restoring forces equal to or greater than the disturbing forces to maintain the state of equilibrium. Stability problem is concerned with the behavior of the synchronous machines after a disturbance. For convenience of analysis, stability problems are generally divided into two major categories steady state stability and transient stability. Steady state stability is the ability of the power system to regain synchronism after small and slow disturbance, such as gradual power changes. An extension of the steady state stability is known as the dynamic stability which is concerned with small disturbances lasting for a long time with the inclusion of automatic control devices. Transient stability studies deal with the effect of large, sudden disturbances, such as the occurrence of a fault, the sudden outage of a line or the sudden application or removal of loads [4]. Since, the transient stability has a role to play in case of major disturbances, the improvement of this stability is important. Many researchers and power system engineers have worked and proposed different methods and techniques to control the stability problem [5-8]. One of these technologies is Flexible AC Transmission System (FACTS) that is a family of several power electronic products [9-11].

FACTS devices are essential to change the power system parameters in order to obtain a better system operation in a faster and more effective way [12]. The voltage source converter based series compensator, called Static Synchronous Series Compensator (SSSC) is a part of FACTS devices family which is connected in series with the power system and was firstly proposed in 1989 [13]. SSSC provides the virtual compensation of transmission line impedance by injecting the controllable voltage in series with the transmission line. The virtual reactance inserted by the injected voltage source influences electric power flow in the transmission lines independent of the magnitude of the line current. The ability of SSSC to operate in capacitive as well as inductive mode makes it very effective in controlling the power flow of the system. Apart from the stable operation of the system with bidirectional power flows, the SSSC has an excellent response time and the transition from positive to negative power flow through zero voltage injection is perfectly smooth and continuous [14-16]. Moreover, the reactive power or current of SSSC can be adjusted by controlling the magnitude and phase angle of the output voltage of the shunt converter [17-18]. The contribution of this paper lies in handling of electrical power flow and stability by connecting SSSC in the network.

2.   Research Method

The input mechanical power from the prime mover of the generator, P_m is assumed as a constant quantity and the electrical power output, P_e will determine whether the machine be in synchronism or not indicating whether the rotor decelerates, accelerates, or remains at the synchronous speed. Until P_e equals P_m the machine maintains synchronism. But if P_e changes from P_m, the rotor consequently deviates from the steady state synchronous speed. Any electrical disturbance due to faults, sudden load change or circuit breaker operations would change the electrical output, P_e in a rapid way resulting in a transient stability problem. This change in P_e depends on the transmission and distribution network and also on the loads to which the machine supplies power [19].

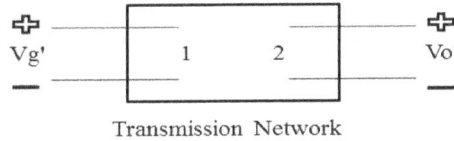

Transmission Network

Figure 1. Schematic Diagram for Stability Studies

Figure 1 shows the reduced schematic diagram of a single machine infinite bus system where the generator at bus 1 is supplying power to the receiving end at bus 2 through the transmission system consisting of the various passive components including transmission line, transformers, circuit breakers, capacitors and the transient reactance of the generators. V_g' represents the transient internal voltage of generator at bus 1 and V_o' at the receiving end is regarded as that of an infinite bus or as the transient internal voltage of a synchronous motor. Now, the bus admittance matrix of the network can be considered as follows reducing to two nodes.

$$Y_{bus} = \begin{bmatrix} Y_{11} & Y_{12} \\ Y_{21} & Y_{22} \end{bmatrix} \tag{1}$$

And for the complex power of any bus,

$$P_k - jQ_k = V_k^* \sum_{n=1}^{N} Y_{kn} V_n \tag{2}$$

In this case, K=1 and N=2 therefore,

$$P_1 + jQ_1 = V_g' (Y_{11} V_g')^* + V_g' (Y_{12} V_0')^* \tag{3}$$

Where, the parameters are defined as:

$$V_g' = |V_g'| \angle \delta_1 \quad ; \quad V_0 = |V_0'| \angle \delta_2 \quad and$$
$$Y_{11} = G_{11} + jB_{11} \quad ; \quad Y_{12} = |Y_{12}| \angle \theta_{12}$$

Now, from "(3)",

$$P_1 = \left|V_g'\right|^2 G_{11} + \left|V_g'\right|\left|V_0'\right|\left|Y_{12}\right| \quad \cos\ (\delta_1 - \delta_2 - \theta_{12}) \tag{4}$$

$$Q_1 = -\left|V_g'\right|^2 B_{11} + \left|V_g'\right|\left|V_0'\right|\left|Y_{12}\right| \quad \sin\ (\delta_1 - \delta_2 - \theta_{12}) \tag{5}$$

If two new angles are defined as,

$$\delta = (\delta_1 - \delta_2) \quad and \quad \gamma = (\theta_{12} - \frac{\pi}{2})$$

"(4)", and "(5)", yield that:

$$P_1 = \left|V_g'\right|^2 G_{11} + \left|V_g'\right|\left|V_0'\right|\left|Y_{12}\right| \sin\ (\delta - \gamma) \tag{6}$$

$$Q_1 = -\left|V_g'\right|^2 B_{11} + \left|V_g'\right|\left|V_0'\right|\left|Y_{12}\right| \cos\ (\delta - \gamma) \tag{7}$$

"(6)", can be rewritten in the following form,

$$P_e = P_c + P_{max} \quad \sin\ (\delta - \gamma) \tag{8}$$

Where,

$$P_c = \left|V_g'\right|^2 G_{11} \quad and \quad P_{max} = \left|V_g'\right|\left|V_0'\right|\left|Y_{12}\right|$$

Neglecting the armature copper loss, P_c "(8)," can be written as:

$$P_e = P_{max} \quad \sin\ \delta \tag{9}$$

Where, $P_{max} = \dfrac{\left|V_g'\right|\left|V_0'\right|}{X}$

X is the transfer reactance between the voltages $V^{g'}$ and $V^{o'}$. "(9)", is known as the power angle equation.

A Single Machine Infinite Bus (SMIB) system is considered as shown in the Figure 2 with its equivalent single line diagram illustrated in Figure 3. The generator is being represented by a constant voltage source V_g' with a transient reactance X_d. And X_1, X_2, X_3 are the equivalent reactance between bus 1 and 2, bus 2 and 3 and bus 3 and 4 respectively whereas bus 4 is considered as the infinite bus. The output electrical power of the SMIB transmission network can be evaluated by the power angle Equation "(9)".

$$P_e = \frac{\left|V_g'\right|\left|V_0'\right|}{X} \sin\ \delta \tag{10}$$

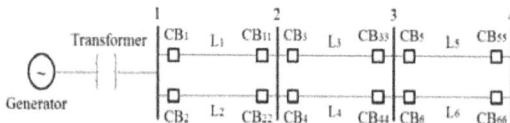

Figure 2. Schematic Diagram of SMIB

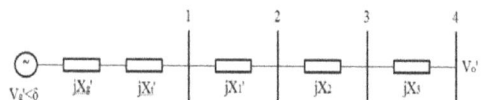

Figure 3. Single Line Diagram of SMIB

In order to control three parameters of line power flow such as line impedance, voltage and phase angle, a Static Synchronous Series Compensator (SSSC) that is a Voltage Source Inverter (VSI) is connected in series to the Single Machine Infinite Bus (SMIB) system through a series transformer and this can be seen from Figure 4. In practice, SSSC is connected by a common dc link including storage capacitor [20]. The series inverter can be applied to control the real and reactive line power flow and voltage with controllable magnitude and phase in series with the transmission line. Therefore, the SSSC can fulfill the responsibility of active and reactive series compensation and phase shifting. For this case, the output electrical power may be expressed as:

$$P_e^{ss} = P_e + \Delta P_e^{ss} \qquad\qquad (11)$$

For the capacitive mode V_{ss} is positive and the SSSC supplies a reactive power to the system.

With this mode ΔP_e^{ss} $\Delta P_e^{ss} = \dfrac{V_g' V_{ss}}{X} \sin(\delta - \theta_{ss})$ $\qquad\qquad (12)$

$$\theta_{ss} = \sin - 1\left(\frac{X_3}{V_s V_o}\right)$$

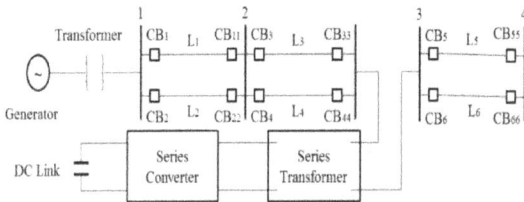

Figure 4. Schematic Diagram of SMIB with SSSC

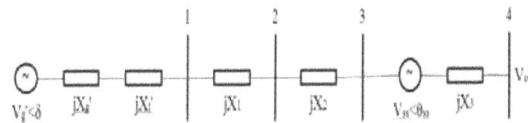

Figure 5. Single Line Diagram of SMIB with SSSC

3. Results and Analysis
3.1. Power Angle Curve

The curve of power angle equation as a function of δ is called power angle curve. Certain parameters are assumed deliberately to with a view to drawing a power angle curve of "Figure 2", such as delivered power of the machine, $P_m=1$ per unit, transient internal voltage, $V_g'=1.05$ per unit, transient reactance of generator, $X_g'=0.2$, transient reactance of transformer, $X_t'=0.1$, line reactance are $X_{L1}= X_{L2}=0.4$, $X_{L3}= X_{L4}=0.5$ and $X_{L5}= X_{L6}=0.3$, infinite bus voltage is $V_o'=1$ per unit and constant $H=5$MJ/MVA.

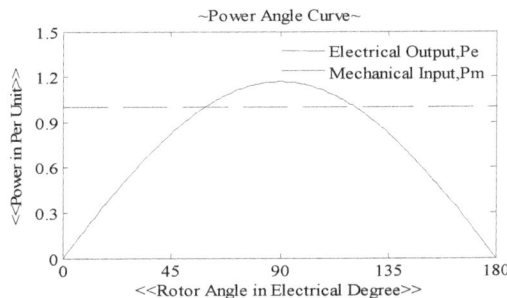

Figure 6. Power Angle Curve at Sound Condition

If it is considered that a three phase fault occurs on line, L_6. In this case, the output power drops to zero and in order to regain the sound operation SMIB system, it is required to open the faulted line L_6. This can be done by opening the circuit breakers. The clearing time of fault is 86ms that can be calculated from "(11)".

$$t_{cr} = \frac{\sqrt{4H(\delta_{cr} - \delta)}}{w\,P_m} \qquad (13)$$

$$\delta_{cr} = \cos - [(\pi - 2\delta_0)\sin\delta_0 - \cos\delta_0]$$

and $\delta_0 = \sin - 1(\frac{P_m}{P_{max}})$

~Power Angle Curve after Fault Clearance~

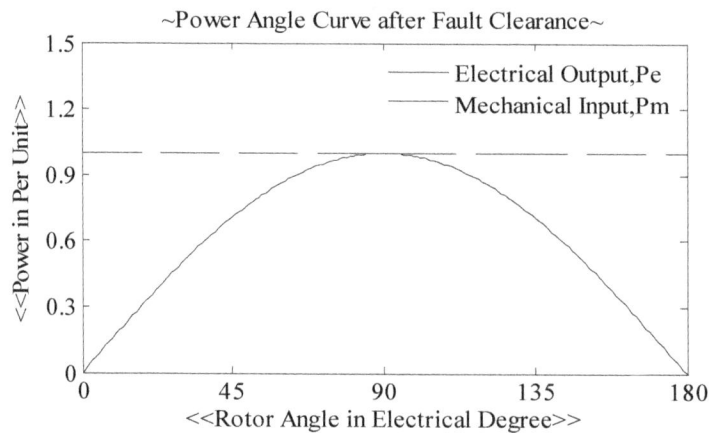

Figure 7. Power Angle Curve at Post Fault Condition

According to the expectation, after clearing fault, the electrical output power has been increased demonstrated in Figure 7 but this value is less than the output power if there is no fault.

3.2. Transient Stability Improvement

SSSC has a secondary function but important that is used to control stability for standing power system oscillations and therefore, SSSC has the capability of improving the transient stability of power system. The transient stability of SMIB system can be expressed by the swing Equation "(14)", which describes the acceleration or de-acceleration of rotor with synchronously rotating air gap mmf during any disturbance and control strategy equation of SSSC "(15)",

$$\frac{d^2\delta}{d^2 t} = \frac{w_s}{2H}(P_m - P_e) \qquad (14)$$

$$w_s = \frac{d\delta}{dt} \quad and \quad V_{ss} = K\,w_s \qquad (15)$$

Where, K is a constant gain control specified as 100.

A curve that contains the solution of swing equation for the expression of δ as a function of time is called swing curve. It helps to determine whether the system remains in synchronism after the occurrence of disturbance or not [19]. As far as Figure 8, representing swing curve is concerned, it is evident that SMIB system is unstable and that can be made stable only a little time later if SSSC is inserted in it.

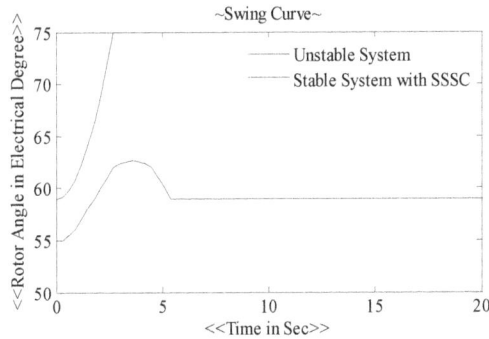

Figure 8. Transient Stability of SMIB System with and without SSSC

4. Conclusion

The proposed strategy of using a Static Synchronous Series Compensator (SSSC) in a Single Machine Infinite Bus (SMIB) system furnishes knowledge that through it, not only electrical power flow can be influenced but also stable operation can be achieved.

As the days are going, the power systems are becoming modernized requiring flexibility, high flow of power and stable operation. Because of having the capability of providing flexibility and rapidness of power flow in the power system network, the utilization of SSSC is expected to grow more and there is a great possibility that new realistic models of this controller will be appeared in the upcoming days.

References

[1] P Kumratug. Improving Power System Transient Stability with Static Synchronous Series Compensator. *American Journal of Applied Sciences*. 2011; 8(1): 77-81.
[2] GA Ajenikoko et al. A Model for Assesment of Transient Stability of Electrical Power System. *International Journal of Electrical and Computer Engineering (IJECE)*. 2014; 4(4): 498-511.
[3] P Thannimalai et al. Voltage Stability Analysis and Stability Improvement of Power System. *International Journal of Electrical and Computer Engineering (IJECE)*. 2015; 5(2): 189-197.
[4] Hadi Sadat. Power System Analysis. Newyork: Mcgraw-Hill Companies, Inc. 2002.
[5] N. M. Al-Rawi et al. Computer Aided Transient Stability Analysis. *Journal of Computer Science*. 2007; 3: 149-153.
[6] S Babainejad et al. Analysis of Transient Voltage Stability of a Variable Speed Wind Turbine with Doubly Fed Induction Generator Affected by Different Electrical Parameters of Induction Generator. *Applied Sci. Res.* 2010; 5: 251-278.
[7] AK Sahoo et al. Power Flow Study Including FACTS Devices. *American. Journal of Applied Sciences*, 2010; 10: 1563-1571.
[8] M Parry et al. Adaptive Data Transmission in Multimedia Networks. *American. Journal of Applied Sciences*. 2005; 2: 730-733.
[9] MA Hannan et al. Transient Analysis of FACTS and Custom Power Devices Using Phasor Dynamics. *J. Applied Sci.* 2006; 6: 1074-1081.
[10] S Chettih et al. ptimal Distribution of the Reactive Power and Voltages Control in Algerian Network Using the Genetic Algorithm Method. *Inform. Technol. J.* 2008; 7: 1170.1175.
[11] IA Muwaffaq. Derivation of UPFC DC Load Flow Model with Examples of its Use in Restructured Power Systems. *IEEE Trans. Power System*. 2003; 18: 1173-1180.
[12] S Teerathana et al. *An Optimal Power Flow Control Method of Power System by Interline Power Flow Controller (IPFC)*. Proceeding 7th International Power Engineering Conference. *Singapore*. 2005: 1-6.
[13] L Gyugyi. *Solid-State Control of Electric Power in ac Transmission Systems*. International Symposium on Electric Energy Conversion in Power Systems. Capn, Italy. 1989.
[14] S C Swain et al. Design of Static Synchronous Series Compensator Based Damping Controller Employing Real Coded Genetic Algorithm. *International Journal of Electrical and Electronics*. 2011; 5 (3).
[15] L Gyugyi. Dynamic Compensation of ac Transmission Lines by Solid-State Synchronous Voltage Sources. *IEEE Trans. Power Delivery*. 1994; 9(2): 904–911.
[16] L Gyugyi et al. Static Synchronous Series Compensator: a Solid State Approach to the Series Compensation to Transmission Lines. *IEEE Trans. Power Delivery*. 1997; 12(1): 406-417.

[17] AT Seyed et al. Decentralized Controller Design for Static Synchronous Compensator Using Robust Quantitative Feedback Theory Method. *American. Journal of Applied Sciences*. 2009; 1: 66-75.

[18] AN Al-Husban. An Eigenstructure Assignment for a Static Synchronous Compensator. *American Journal of Applied Sciences*. 2009; 2: 812-816.

[19] WD Stevenson Jr. Elements of Power System Analysis. Fourth Edition. McGraw-Hill Book Company. 2011-2012.

[20] CR Fuerte-Esquivel et al. A Comprehensive Newton-Raphson UPFC Model for the Quadratic Power Flow Solution of Practical Power Networks. *IEEE Trans. Power System*. 2000; 15(1): 102-109.

13

A Document Imaging Technique for Implementing Electronic Loan Approval Process

J. Manikandan, C.S. Celin, V.M. Gayathri
Department Of Computer Science and Engineering,
Saveetha School of Engineering, Saveetha University, Chennai
e-mail: manikandancse8@gmail.com, celin.cs7@gmail.com, vmg188@gmail.com

Abstract
The image processing is one of the leading technologies of computer applications. Image processing is a type of signal processing, the input for image processor is an image or video frame and the output will be an image or subset of image [1]. Computer graphics and computer vision process uses an image processing techniques. Image processing systems are used in various environments like medical fields, computer-aided design (CAD), research fields, crime investigation fields and military fields. In this paper, we proposed a document image processing technique, for establishing electronic loan approval process (E-LAP) [2]. Loan approval process has been tedious process, the E-LAP system attempts to reduce the complexity of loan approval process. Customers have to login to fill the loan application form online with all details and submit the form. The loan department then processes the submitted form and then sends an acknowledgement mail via the E-LAP to the requested customer with the details about list of documents required for the loan approval process [3]. The approaching customer can upload the scanned copies of all required documents. All this interaction between customer and bank take place using an E-LAP system.

Keywords: image processing, document imaging, E-LAP system

1. Introduction

These documents include many common types: business letters, forms, engineering drawing & maps, text books, technical manual, music notations & other symbolic data. Though paper was the limited medium in history, many documents now begin on the computers & often dwell entirely in electronic form. In spite of this it is ambiguous whether the computer has decreased/increased the amount of paper document formed, as these are printed out for reading, distribution, markup predictions of paperless offices made so frequently during the early 1980 has given way to the recognition that the objective is not eradication of paper but the ability to deal with the flow of electronic & paper document in effective & incorporated way. Document processing in any association whether having its operations manual or automated, forms an important activity in its functioning's. Within document processing, the key activity prior to all other actions is the identification of documents and hence their classification [1, 7] [10]. Several superior solutions are present for document processing and analysis, this paper tries to give general idea for document processing and the various steps/methods used for that.

Loan processing is one of the tedious process in the banking and financial industries. A loan approval system (E-LAP) is developed specially to support banks and other financial industries loan claim processing needs. It employ workflow technology to control and monitor the various steps in loan processing and uses digital imaging technology to reduce the delays and overhead associated with paper documents. This paper explains about enabling company's loan department to process the loan applications in a structured manner and helps in improving the response time and overall throughput time of loan processing process. The proposed system automates the loan approval process by enabling online requests, modification and it's tracking [4].

The proposed web-based application will be accessed by the applicants and loan officers over the internet. The loan applicants will be able to apply for a loan and track responses and action items. Whereas the loan team at the finance company will be able to view, respond and process loan applications. E-LAP will provide a secure user name and secret code based secured login mechanism to access its services. The proposed E-LAP system will deliver: Easy online loan application access for loan seekers. Better control over turnaround time for loan application processing. Many financial industries have a test to manage competition. Aspects like customer service, relationship, short turnaround time have become critical factors, Data security, real time availability are some of the critical to success factors.

2. Document Image Processing

The document image processing techniques has become very familiar for information retrieval process in past years. The objective of Document Image analysis is to recognize the text & graphics components in image of documents & to extract intended information from them. Two categories of document image analysis can be defined. First one is Text Processing it Deals with the textual components of a document image. The task of text processing is to determine the skew in the document and identifying relevant attributes such as font, size in documents. The second category is graphical processing it deals with non-textual elements such as symbols, images, logos.

The paper documents are now can be converted into electronic documents for easy storage and better processing of information [6]. The document imaging processing is used to attain a document with a high value, precision and fast recovery. Document image processing technique involves extracting the exact content in a computerized method. This document image processing is an important mission for many organizations for many applications. Document classification is a vital mission in document Processing. Document classification permits the automatic allocation or archiving of documents. For instance, after classification of Business letters according to sender and message type the letters are sent to the concern departments for processing. Document classification improves indexing efficiency in Digital Library construction. The document image processing includes three following stages: pre-processing, feature mining and sorting. In pre-processing stage includes image attainment, Linearization, detection, layout processing. In feature mining stage, the important features in documents are preserved and additional features are included. In categorization stage performs office computerization, digital libraries and other document image processing applications [7]. The steps which are used to process the documents are shown in Figure 1. The objective of pre-processing step is to improve the quality of the document images and find the required content from the data. The feature mining step is used to identify the unique features of the document. In categorization step the common features are grouped for easy identification and retrieval process.

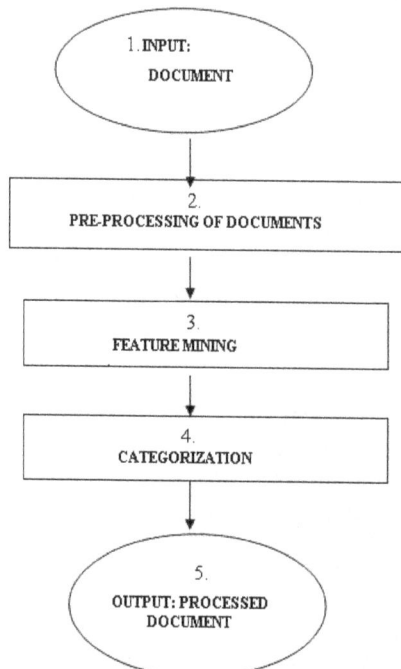

Figure 1. Document Image Processing Steps

The pre-processing step performs removal of unwanted noise in an image, reducing gray scale or color image to binary image, segmentation of different components in image. The given input image is then converted to gray scale or color image or binary formats. Then linearization is performed to differentiate the foreground and background information. In last step de-skewing process is performed. The deviation from the bottom line of page from horizontal direction is called skew. This process involves arranging the paper layout properly, so that it will be ideal for scanning process [16]. The process of de-skewing is shown in Figure 2(a) and Figure 2(b). In feature mining step, only the significant information are extracted from the given input document image. There are two types of features namely, local features and global features. The features extracted from particular block or subsection of image is known as local image. The feature extracted from the entire image is known as global image. Geometric, textual, content and structural based are some of the types of feature mining. One of the main benefits of feature mining is that it highlights only the significant information in an image. Finally in categorization technique, the similar information in images are grouped, so it reduces the number of searches required to find the documents and percentage of error rate in the documents, easy reorganization of documents. The categorization process main role in document image processing and it enhance the document image analysis and indexing features of the documents.

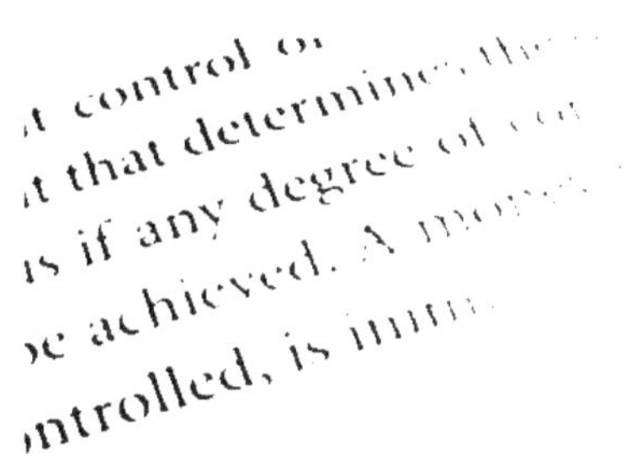

Figure 2(a). Document Imaging Process Flow 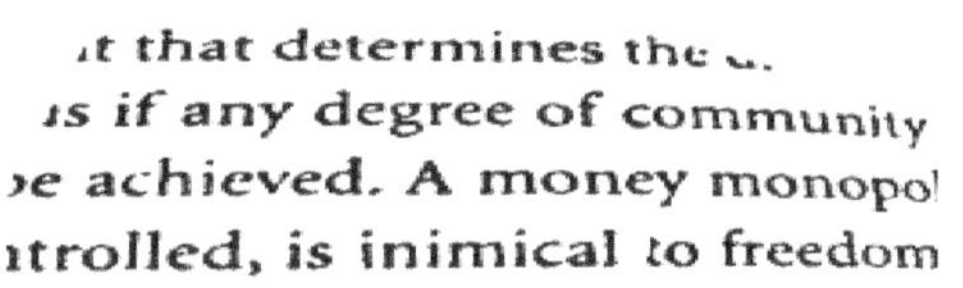   Figure 2(b). Document Imaging Process Flow

The applicants visits and submit the scanned documents on the banking website. The web server collects all these documents and stores in document imaging database. Then the collected documents are prepared according to the banking terms and conditions [9]. Then again the prepared documents are scanned for storage. Before storing the scanned documents, the documents are indexed for easy querying and retrieval processes. The indexing can be done manual and automatic indexing technique. Finally the indexed scanned documents are stored in the document imaging databases. By sending appropriate query to document imaging database we can view and retrieve the documents for processing. The process flow of document imaging process is shown in Figure 3.

3. Related Work
Loan processing system discusses about an abstract decision-making model that evaluates a loan file in the same way that a normal human process. Risk management approval system is used to predict the future and used to avoid the risks as much as possible. Loan selection emphasizes on the identification and management of risk among groups of loans applied [11]. The credit evaluation system checks the applicability of one of the new integrated model on a sample data taken from banks. Prior relationships in bank loan analyses the importance of retail customer's banking relationships for loan defaults using a distinctive, broad dataset of over one million loans by banks. Single loan opening system discusses the development of a comprehensive dataset of over one million loans by saving banks [13]. Single loan initiation system discusses the developments of a wide-ranging and flexible loan originating system that can be readily and seamlessly interfaced with a core banking system to cater to the banks end to end transaction processing.

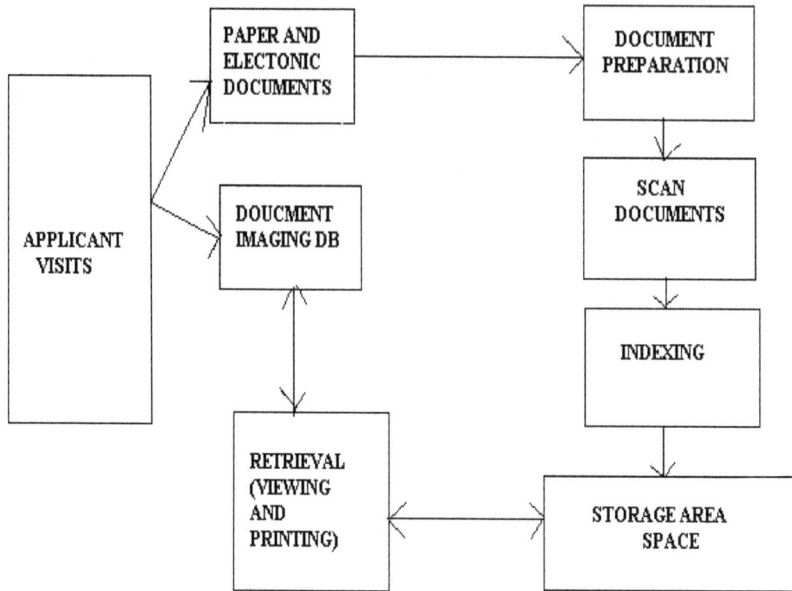

Figure 3. Document Imaging Process Flow

Mind tree uses the business process modeling technology that promises to automate the corporate loan originating process, while retaining partisanship in decision making in the loan approval process suggest solutions that could improve the customer experience in the approving process loan assessment suggests in reviewing the existing literature on decision support system for loan assessment in financial companies [15]. The loan approval process model focuses on the modeling of business process that a banking society follows, during the actions of a loan request.

4. Proposed Work

There has been many research works in the past three years in implementing web-service based bank application systems. This paper focuses on the optimization of the approval process a bank conducts during the application process of small business loans. An ideal system should have the following. Have a structured workflow for automatic routing of applications for different loan products, support multiple loan products and loan types, support, different origination channels, easily interface with different origination channels, easily interface with different external, provide internal credit inspection, credit calculations and conformity examination, preserve all the request details and provide status updates as and when required, support document imaging and archiving for various loan application related documents. The proposed web-based application, E-LAP will be used by the applicants and loan officers over the internet. The loan applicants shall be able to apply for a home loan and track responses and action items. Whereas the loan team at the bank shall be able to view, respond, and process loan applications. The admin user will be able to upload masters in to the application.

The city master will be uploaded from the CSV (comma separated values) format file. The CSV has one column (city). No duplicate records in this file. The checklist of documents required per city for home loan processing. The CSV file has three columns that is city, name of the document, number of copies. Such that this will be received from loan department in a CSV format and will be periodically uploaded by the E-LAP admin user. The E-LAP database used to hold all the E-LAP data including the uploaded data loan applications, processing status, uploaded documents. The web server is used to run the bank website, which contains the E-LAP link. The mail server used for providing acknowledgement to the applicants after the loan is applied for and is being processed by the loan department. The system architecture of E-LAP system is shown in Figure 4.

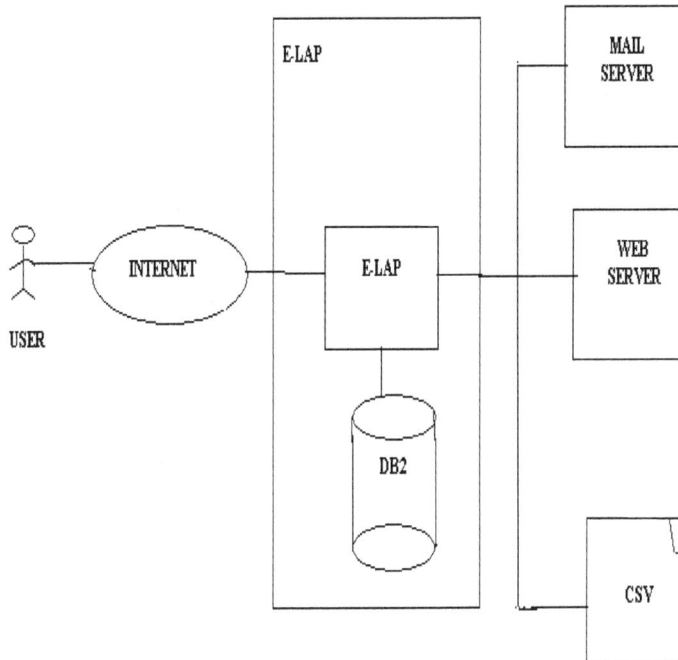

Figure 4. E-LAP System Architecutre

5. Conclusion

A loan approval system that is seamlessly integrated with the core banking solution is becoming a critical requirement for banks. Thus to develop such a system that optimizes the time and space complexities of the existing system in the banking sector is highly essential. This system also uses digital imaging technology to reduce the delays and inefficiencies associated with paper documents. The home loan processing complexity is toned down for purpose of project timelines. Automatic loan EMI or interest calculations are not required; the developer will treat them as data entry fields. Most of the concerns needs to store and query huge amount of information in electronic format for retrieving effectively and easily. Thus document image processing plays a vital role in various applications like banking and financial institutions.

References

[1] Hedjam R, Cheriet M. *Ground-Truth Estimation in Multispectral Representation Space: Application to Degraded Document Image Binarization.* Document Analysis and Recognition (ICDAR), 2013 12th International Conference. 190-194.
[2] Obafemi-Ajayi T. Agam G. Character-Based Automated Human Perception Quality Assessment in Document Images. *Systems, Man and Cybernetics, Part A: Systems and Humans, IEEE Transactions on.* 2012; 42(3): 584,595.
[3] Heatwole E. Processing document images on the telco network. *Communications Magazine, IEEE.* 1993; 31(1): 40-44.
[4] Wu XD, Street RA, Weisfield R, Ready S, Nelson S. *Page sized a-Si:H 2-dimensional array as imaging devices.* Solid-State and Integrated Circuit Technology, 4th International Conference on. 1995: 724,726.
[5] Nunnagoppula G, Deepak KS, Rai HGN, Krishna PR, Vesdapunt N. *Automatic blur detection in mobile captured document images: Towards quality check in mobile based document imaging applications.* Image Information Processing (ICIIP), IEEE Second International Conference on. 2013: 299-304.
[6] Yun Lin, Seales WB. *Opaque document imaging: building images of inaccessible texts.* Computer Vision, ICCV 2005. Tenth IEEE International Conference on. 2005; 1: 662-669.
[7] Lemmi F, Mulato M, Ho J, Lau R, Lu JP, Street RA, Palma F. Active matrix of amorphous silicon multijunction color sensors for document imaging. *Applied Physics Letters.* 2001; 78(10): 1334-1336.

[8] Barbu E, Heroux P, Adam S, Trupin E. Clustering document images using a bag of symbols representation. *Document Analysis and Recognition. Proceedings. Eighth International Conference on.* 2005; 2: 1216-1220.

[9] Bolan Su, Shuangxuan Tian, Shijian Lu, Thien Anh Dinh, Chew Lim Tan. *Self Learning Classification for Degraded Document Images by Sparse Representation.* Document Analysis and Recognition (ICDAR), 12th International Conference on. 2013: 155-159.

[10] Shijian Lu, Linlin Li, Tan CL. Document Image Retrieval through Word Shape Coding. *Pattern Analysis and Machine Intelligence, IEEE Transactions on.* 2008; 30(11): 1913-1918.

[11] Bolan Su, Shijian Lu, Chew Lim Tan. Robust Document Image Binarization Technique for Degraded Document Images. *Image Processing, IEEE Transactions on.* 2013; 22(4): 1408-1417.

[12] Trahanias, PE, Venetsanopoulos AN. Vector directional filters-a new class of multichannel image processing filters. *Image Processing, IEEE Transactions on.* 1993; 2(4): 528-534.

[13] Prajapati HB, Vij SK. *Analytical study of parallel and distributed image processing.* Image Information Processing (ICIIP), 2011 International Conference on. 2011: 1-6.

[14] Chien SA. *Automated synthesis of image processing procedures for a large-scale image database.* Image Processing, 1994. Proceedings. ICIP-94., IEEE International Conference. 1994; 3: 796-800.

[15] Yi Liang, Yingyuan Xiao, Jing Huang. *An Efficient Image Processing Method Based on Web Services for Mobile Devices.* Image and Signal Processing, 2009. CISP '09. 2nd International Congress on. 2009: 1-5.

[16] Casey RG, Wong KY. Document Analysis Systems and Techniques, Image Analysis Applications. *Image Analysis Applications.* 1990: 1-35.

[17] Castleman KR. Digital Image Processing. Englewood Cliffs, NJ: Prentice-Hall, Inc. 1979.

[18] CC Chang, DC Lin s. A Spatial DataRepresentation: an Adaptive 2D-H string. *Pattern Recognition Letters 17.* 1996: 175-185.

[19] Sonka Milan, Hlavac, Roger Boyle. Image Processing Analysis and Machine Vision. Brooks/Cole Thomson Learning. 1999.

[20] T Young, Gerbrands. Fundamentals of Image Processing. Paper Back. 2007.

[21] O Gorman L, Kasturi R. Document Image Analysis Systems. *Computer.* July 1992; 25: 5-8.

[22] Rangachar Kasturi1, Lawrence, O'gorman2. Document image analysis: A primer. S⁻adhan⁻a; 2002: 3-22.

14

Improved Rotor Speed Brushless DC Motor using Fuzzy Controller

Jafar Mostafapour[1]*, Jafar Reshadat[1,3], Murtaza Farsadi[2]
[1]Azerbaijan Regional Electric company, Tariz, Iran
[2]Department of Electrical Engineering, Urmia University
[3]Department Management, Science and Research Branch Islamic Azad University, West Azerbaijan, Iran
e-mail: j.mostafapour.a@gmail.com

Abstract

A brushless DC (BLDC) Motors have advantages over brushed, Direct current (DC) Motors and, Induction motor (IM). They have better speed verses torque characteristics, high dynamic response, high efficiency, long operating life, noiseless operation, higher speed ranges, and rugged construction. Also, torque delivered to motor size is higher, making it useful in application where space and weight are critical factors. With these advantages BLDC motors find wide spread application in automotive appliance, aerospace medical, and instrumentation and automation industries This paper can be seen as fuzzy controllers compared to PI control BLDC motor rotor speed has improved significantly and beter result can be achieve

Keywords: BLDC, matlab/simulink, PID controller, PID fuzzy controller

1. Introduction

Brushless motor technology makes it possible to achieve high reliability with high efficiency, and for a lower cost in comparison with brush motors. Although the brushless characteristic can be apply to several kinds of motors AC synchronous motors, stepper motors, switched reluctance motors, AC induction motors the BLDC motor is conventionally defined as a permanent magnet synchronous motor with a trapezoidal Back, Electric Magnetic Fields (EMF) waveform shape. Permanent magnet synchronous machines with trapezoidal Back-EMF and (120 electrical degrees wide) rectangular stator currents are widely used as they offer the following advantages first, assuming the motor has pure trapezoidal Back EMF and that the stator phases commutation process is accurate, the mechanical torque developed by the motor is constant secondly, the Brushless DC drives show a very high mechanical power density. Brushless Direct Current (BLDC) motors are one of the motor types rapidly gaining popularity. BLDC motors are used in industries such as Appliances, Automotive, Aerospace, Consumer, Medical, Industrial Automation Equipment and instrumentation [1, 2], As the name implies, BLDC motors do not use brushes for commutation; instead, they are electronically commutated. BLDC motors have many advantages over brushed DC motors and induction motors

BLDC Motors are available in many different power ratings, from very small motors as used in hard disk drives to larger motors used in electric vehicles. Purpose of this article IS improve the performance of BLDC rotor speed By using fuzzy control and compared with PI controller.

2. Working of BLDC Motor

The BLDC motor is an AC synchronous motor with permanent magnets on the rotor (moving part) and windings on the stator (6 part). Permanent magnets create the rotor flux and the energized stator windings create electromagnet poles. The rotor (equivalent to a bar magnet) is attracted by the energized stator phase. By using the appropriate sequence to supply the stator phases, a rotating field on the stator is created and maintained. This action of the rotor - chasing after the electromagnet poles on the stator is the fundamental action used in synchronous permanent magnet motors [2]. The lead between the rotor and the rotating field must be controlled to produce torque and this synchronization implies knowledge of the rotor position.

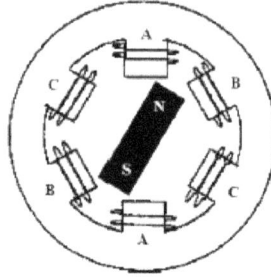

Figure 1. A three-phase synchronous motor with a one permanent magnet pair pole rotor

On the stator side, stator is three phase similar to induction motor These offer a good compromise between precise control and the number of power electronic devices required to control the stator currents. For the rotor, a greater number of poles usually create a greater torque for the same level of current. On the other hand, by adding more magnets, a point is reached where, because of the space needed between magnets, the torque no longer increases. The manufacturing cost also increases with the number of poles. As a consequence, the number of poles is a compromise between cost, torque and volume [3]. Permanent magnet synchronous motors can be classified in many ways, one of these that is of particular interest to us is that depending on back EMF profiles: Brushless Direct Current Motor (BLDC) and Permanent Magnet Synchronous Motor (PMSM). This terminology defines the shape of the back EMF of the synchronous motor. Both BLDC and PMSM motors have permanent magnets on the rotor but differ in the flux distributions and back EMF profiles.

To get the best performance out of the synchronous motor, it is important to identify the type of motor in order to apply the most appropriate type of control is described. We have seen that the principle of the BLDC motor is, at all times, to energize the phase pair which can produce the highest torque. To optimize this effect the Back EMF shape is trapezoidal. The combination of a DC current with a trapezoidal Back EMF makes it theoretically possible to produce a constant torque. In practice, the current cannot be established instantaneously in a motor phase; as a consequence the torque ripple is present at each 60° degree phase commutation [4, 5].

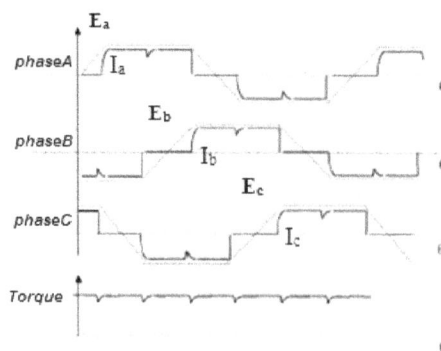

Figure 2. Electrical Waveforms in the Two Phase ON Operation and Torque Ripple

2.1. Mathematical Model of BLDCM

As shown in Figure 3, a dynamic equivalent circuit of the BLDC motor. For this model, the stator phase voltage equations in the stator reference frame of the BLDC Motor are given as in Equation (1, 5) which are provided below. The following assumptions are made:1) the three phase windings are symmetrical, 2) magnetic saturation is neglected, 3) hysteresis and eddy current losses is not considered, and 4) the inherent resistance of each of the motor windings is R ,the self-inductance is L, and the mutual inductance is M.

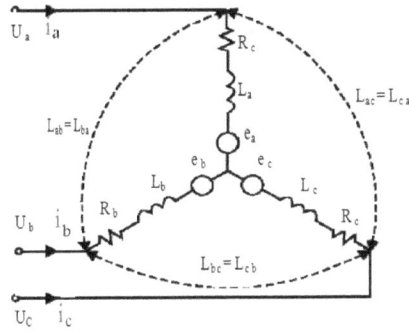

Figure 3. Dynamic equalent circuit

$$\frac{d}{dt}\begin{bmatrix} i_a \\ i_b \\ i_c \end{bmatrix} = \frac{1}{(L-M)}\left\{ \begin{bmatrix} U_a \\ U_b \\ U_c \end{bmatrix} - \begin{bmatrix} R_a & 0 & 0 \\ 0 & R_b & 0 \\ 0 & 0 & R_C \end{bmatrix}\begin{bmatrix} i_a \\ i_b \\ i_c \end{bmatrix} + \omega_r \Psi_m \begin{bmatrix} f_a(\theta) \\ f_b(\theta) \\ f_c(\theta) \end{bmatrix} \right\} \qquad (1)$$

Where, Ua, Ub and Uc are the phase voltage of three-phase windings, ia, ib and ic are the phase current, and ea, eb and ec are the back EMF.

$$f_a(\theta_r) = 1 \qquad 0^\circ \angle \theta_r \angle 180^\circ$$

$$= (\frac{6}{\pi})\ (\pi - \theta_r)\ -1 \quad 120^\circ \angle \theta_r \angle 180^\circ \qquad (2)$$

$$= -1 \quad 180^\circ \angle \theta_r \ \angle 300^\circ$$

$$= (\frac{6}{\pi})(\theta_r - 2\pi)\ +1300^\circ \ \angle \theta_r \angle 360^\circ$$

Electrical power of motor can be calculated using Equation (3).

$$P = e_a i_a + e_b i_b + e_c i_c \qquad (3)$$

Electromagnetic torque can also be expressed as Equation (4). Speed is derived from rotor position Θr as in Equation (5).

$$T_e = j(\frac{2}{p})\frac{d}{dt}\omega_r + B_m(\frac{2}{p})\omega_r + T\ 1 \qquad (4)$$

$$\frac{d}{dt}\omega_r = (\frac{p}{2j})(T_e - B_m(\frac{2}{p})\omega_r + T\ 1) \qquad (5)$$

$$\frac{d}{dt}\theta_r = \omega_r \qquad (6)$$

From the above equations, BLDC motor can be modeled [6].

3. A Review on Utilized Systems
3.1. Implementation of PID fuzzy Controller for BLDC
In this section implementation of Fuzzy Inference System for nonlinear fuzzy PID control is explicated using control system toolbox of Simulink.

As mentioned before a fuzzy inference system maps known inputs to outputs using fuzzy logic. For instance, mapping of a controller can be stated by a three dimensional diagram. This diagram is called control surface. The following figure illustrates a hypothetical control surface.

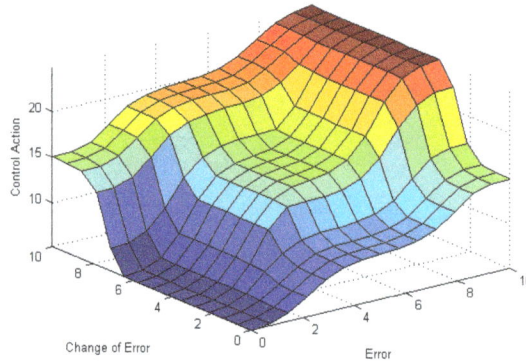

Figure 4. An example of a control surface

Error signal $e(k)$ and error variation signal $(e(k)-e(k-1))$ are common inputs of FIS. The output of FIS is a control operation which is inferred from fuzzy rules.

In our study, utilized system is a BLDC model, single input- single output, which is discretized. The control objective is tracking reference signal.

3.2. Structure of Fuzzy PID Controller
The exploited fuzzy controller is a feedback loop which operates similar to PID which is calculated by fuzzy inference. The closed loop structure in SimuLink is as follows. It can be observed by typing the undergoing instruction.

Open-system ('Fuzzy_PID')

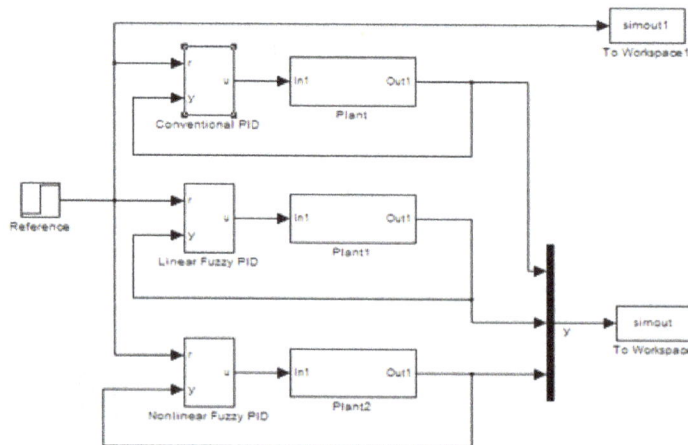

Figure 5. Closed loop structure in SimuLink

Three controllers depicted in the above figure are respectively conventional PID, linear fuzzy PID controller and nonlinear fuzzy PID controller. We will see that it is necessary to design conventional and linear fuzzy PID controllers to design nonlinear fuzzy controller.

Parallel structure is utilized to implement fuzzy controller. It is a combination of fuzzy PI and fuzzy PD controllers. The structure of fuzzy controller is demonstrated in figure below.

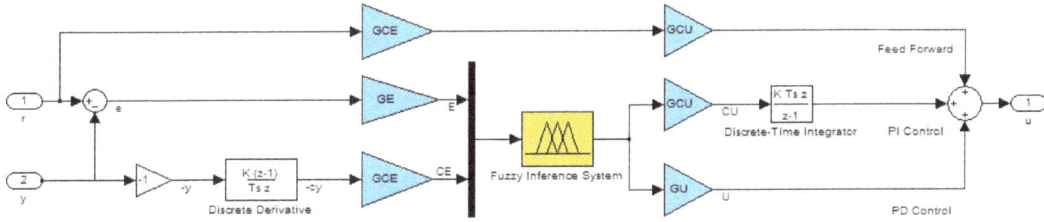

Figure 6. Structure of fuzzy controller

The magnitude of -(y(k)-y(k-1)) is used instead of signal changes. It is done in this way to avoid direct stimulation of derivative signal by step changes in input reference. Two gain blocks, GCE and GCU, are employed in the feedforward path. These two blocks guarantee that error signal e, is used proportional when the fuzzy PI controller is linear

3.3. Design Procedure for Fuzzy PID Controller

Design of fuzzy controller includes configuration of fuzzy inference system and substitution of GE, GCU, GCE and GU scaling factors. Here, the following steps are taken for controller design.

a) Designing conventional PI controller
b) Designing equivalent linear fuzzy PID controller
c) Adjusting fuzzy inference system to obtain nonlinear control surface (designing nonlinear fuzzy PID controller)
d) Optimum adjustment of nonlinear fuzzy PID controller

3.3.1. The First Step: Designing Conventional PID Controller

To implement PID controller, the parallel structure, which is shown below, is exploited.

$$K_p + K_i \frac{T_s z}{z-1} + K_d \frac{z-1}{T_s z} \quad .$$

The mentioned controller is implemented as follows in the SimuLink.

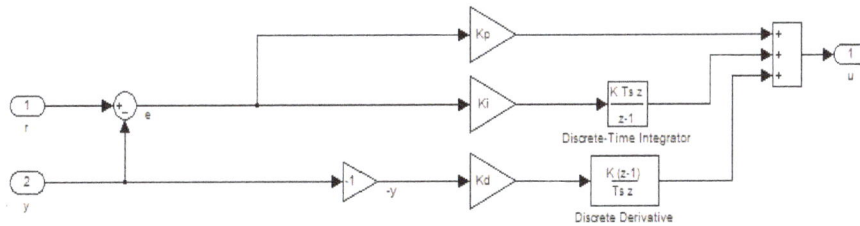

Figure 7. PID structure in the Simulink environment

Similar to fuzzy PID controller the input signal for derivative operator is −y(k).

PID coefficients might be adjusted manually or using adjustment rules. The following instructions might be utilized to adjust PID controller coefficients in control tool box.

```
% Designing Conventional PID
C0 = pid(1,1,1,'Ts',Ts,'IF','B','DF','B'); % define PID structure
C = pidtune (plant,C0); % design PID
[Kp Ki Kd] = piddata(C); % obtain PID gains
```

3.3.2. Second Step: Designing Equivalent Fuzzy PID Controller

With FIS configuration and selecting four scaling coefficients, a fuzzy controller is derived whose performance is exactly the same as conventional PID.

First off, fuzzy system is configured. As a result a linear control surface is achieved from E and CE as inputs to U as output. The structure of utilized inference system is summarized as follows.

a) Mamdani inference system is employed.
b) Algebraic multiplication is used instead of AND.
c) The input range is considered to be [-10,10]
d) The fuzzy sets are triangular and they intersect their neighbors in 0.5 membership value.
e) The output range is [-20, 20].
f) The outputs are single-valued determined by sum of peak positions of input sets.
g) The center of gravity method is used for deffuzification.

Values of input and output ranges and membership function parameters must be assigned so that the relation between input and output of the system is equal to an identity function. In the next section the coefficients of fuzzy PID controller are derived by assuming the identity function for relation of fuzzy inference system.

The following instruction is used to build fuzzy inference system.

```
%Designing Linear Fuzzy Inference System
FIS2 = newfis('FIS2','mamdani','prod','probor','prod','sum');
```

And the fuzzy rules are also defined as follows:

If E is Negative and CE is Negative then u is -20
If E is Negative and CE is Zero then u is -10
If E is Negative and CE is Positive then u is 0
If E is zero and CE is Negative then u is -10
If E is Zero and CE is Zero then u is 0
If E is Zero and CE is Positive then u is 10
If E is Positive and CE is Negative then u is 0
If E is Positive and CE is Zero then u is 10
If E is Positive and CE is Positive then u is 20

Here we utilized fuzzy tool box instructions to create FIS; however, corresponding GUI might be used as well.

The 3d surface is achieved as follows:

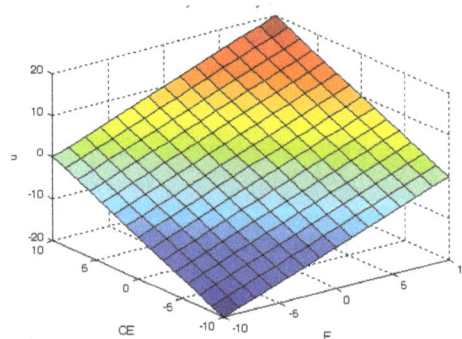

Figure 8. 3D diagram of control surface in fuzzy PID controller

The input and output membership functions are shown below.

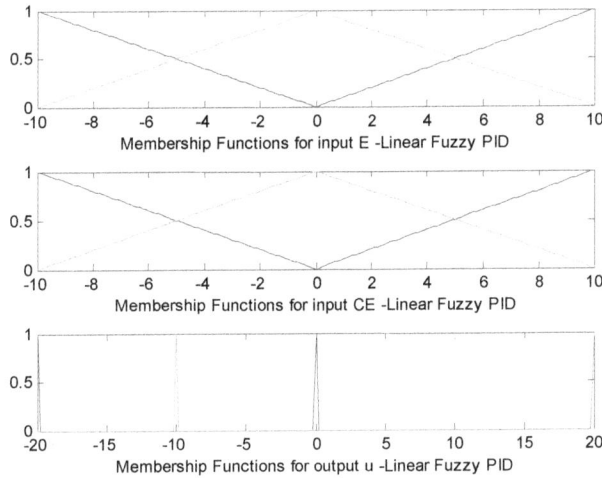

Figure 9. Input and output membership functions for linear fuzzy PID controller

In the next stage, four scaling factors are calculated using coefficients of conventional PID controller. The input-output relation in fuzzy inference system is considered to be in the form of identity function; therefore, the corresponding relations are as shown in the following equations.

$$k_p = GCU \times GCE + GU \times GE$$
$$k_i = GCU \times GE$$
$$k_d = GU \times GCE$$

If the maximum input step is considered as 1 the maximum error value would be 1. Since input range equals to [-10, 10], considering GE=10, GCE, GCU and GU are derived from following equations.

$$GE = 10$$
$$GCSE = GE \times \left. \left(k_p - sqrt\left(k_p^{\,2} - 4k_i k_d \right) \right) \middle/ 2 \times k_i \right.$$
$$GCU = k_i \middle/ GE$$
$$GU = k_d \middle/ GCE$$

The above values are calculated in the corresponding m-file using the following instructions. They are used in Simulink plant file together with controller.

3.3.3. Third Step: Designing Fuzzy PID Controller with Nonlinear Control Surface

First we make sure that fuzzy PID controller is properly designed. Afterwards, FIS adjustments such as, type, functions, membership, fuzzy rules and so on are changed so that desired nonlinear control surface is achieved. For this purpose Sugeno inference system is utilized. Moreover, for each input merely two states, positive and negative, are considered which reduces the number of rules to four.

The fuzzy rule set is defined as follows.

If E is Negative and CE is Negative then u is -20
If E is Negative and CE is Positive then u is 0
If E is Positive and CE is Negative then u is 0
If E is Positive and CE is Positive then u is 20

The 3D diagram of nonlinear control surface is depicted in figure, As shown in Figure 10

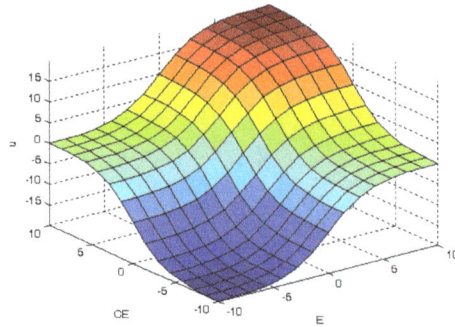

Figure 10. 3D diagram of control surface for nonlinear fuzzy PID controller

As can be seen we have a nonlinear control surface. According to above mentioned control surface, it can be seen that the control surface has considerable gain in the vicinity of center of E and CE plane. As a result when error is small it will decrease more rapidly. When the error is large, the variations of controller are small. It limits control operation and avoids probable saturation. The membership functions for inputs of fuzzy inference system are demonstrated as shown in Figure 11.

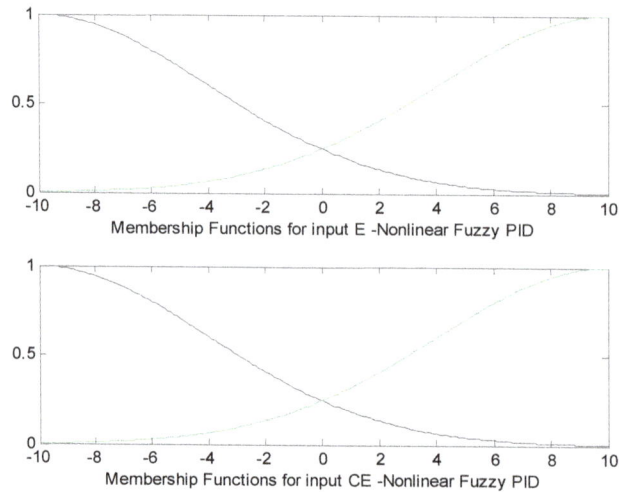

Figure 11. Input and output membership functions associated with nonlinear fuzzy PID controller

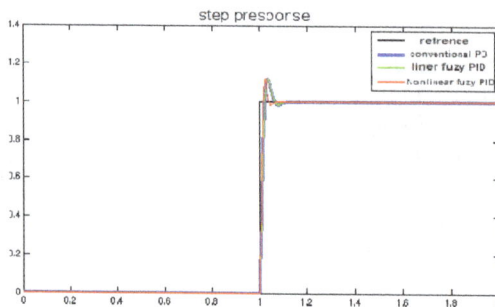

Figure 12. The response of closed loop system with conventional PID, linear fuzzy and nonlinear fuzzy controllers

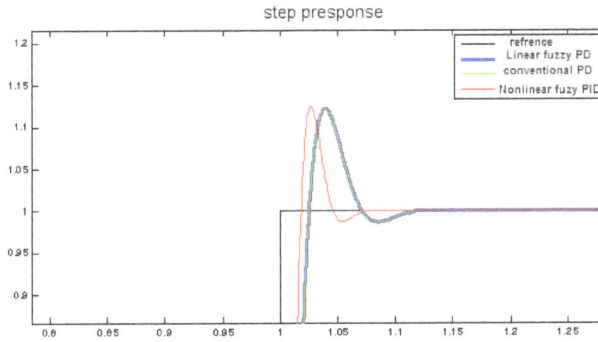

Figure 13. Closed loop system response with conventional PID, linear fuzzy PID and nonlinear fuzzy PID controllers

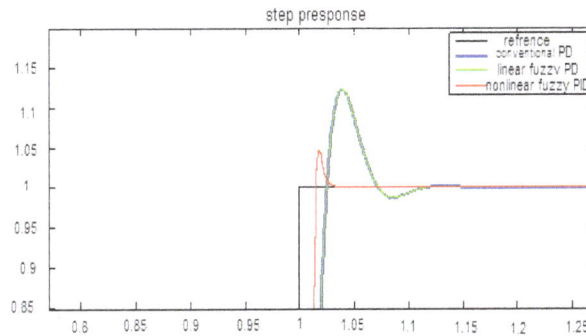

Figure 14. The magnified response of closed loop system with conventional PID, linear fuzzy PID and nonlinear fuzzy PID controllers

As shown in Figure 13, response of system with mentioned controllers is depicted for step input and at t=1s.

As expected the response of the system with conventional PID and linear fuzzy PID are the same. The response of the system with nonlinear fuzzy PID controller is faster than two others; nevertheless, it does not show any improvements regarding overshoot.

3.3.4. Fourth Step: Optimum Adjustment of Nonlinear Fuzzy PID Controller

In this section system response is modified by changing the parameters of input membership functions. The following results are achieved by changing membership functions (changing the parameter related to membership function from 6 to 2)

```
%input E
FIS1 = addvar (FIS1,'input','E',[-10 10]);
FIS1 = addmf (FIS1,'input',1,'Negative','gaussmf',[2 -10]);
FIS1 = addmf (FIS1,'input',1,'Positive','gaussmf',[2 10]);
%input CE
FIS1 = addvar (FIS1,'input','CE',[-10 10]);
FIS1 = addmf (FIS1,'input',2,'Negative','gaussmf',[2 -10]);
FIS1 = addmf (FIS1,'input',2,'Positive','gaussmf',[2 10]);
```

The magnified step response is shown in Figure 16.

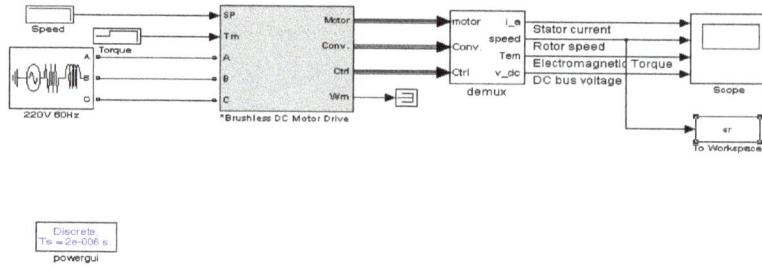

Figure 15. Block Diagram of BLDC

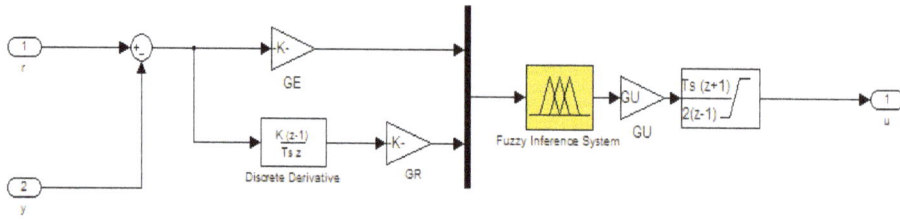

Figure 16. design fuzzy PI controller

As can be seen the response is faster with nonlinear fuzzy PID controller and it has smaller overshoot. Comparing Figure 13 and 14 it can be concluded that the system response is significantly improved by changing the parameters.

4. Simulation and Results

To design fuzzy PI controller, the following structure is used. As we know the structure of a conventional PI is as follows:

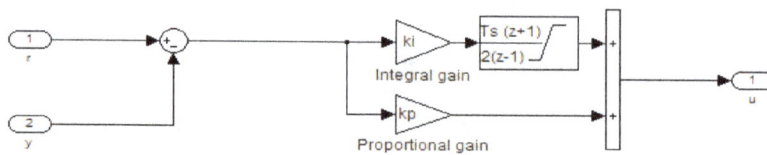

Figure 17. The structure of a conventional PI

Comparing the two structures shows that the two following relationships are established between the coefficients of the controller:

$$ki = GE \times GU$$
$$kp = GR \times GU$$

In the above equations, the coefficients of conventional PI can be obtained by having fuzzy PI controller. In practice, we need to obtain the coefficients of the fuzzy PI controller. For this purpose, by choosing a value for one of the coefficients such as GU, other coefficients can be achieved. The selected values for GU controller determine the degree of nonlinearity for fuzzy controller Here the input range for fuzzy inference system is considered as [-10.10] and for this reason according to the reference signal amplitude and the estimated value that is approximately equal to 300, the value of GU = 1000 is considered that seems to be a good value, an inference system of Sugeno is used as the fuzzy controller. Only two modes of Positive and Negative are considered for each input and Positive, Zero and Negative modes are considered for each output and the total rules have reduced to 4 rules.

To configure the Fuzzy inference system we run the following commands.

If E is Negative and CE is Negative then u is Negative
If E is Negative and CE is Positive then u is Zero
If E is Positive and CE is Negative then u is Zero
If E is Positive and CE is Positive then u is Positive

With the definition of fuzzy inference system and using fuzzy PI controller, the rotor speed's results are obtained as follows:

To compare the fuzzy PI controller and conventional PI controller, a PI with coefficients equal to fuzzy PI is used. Results from both controllers and the reference input are shown in the following figure. Values considered for PI controller are:

$$ki = 46.35, kp = 1.22$$

In the following figure, the system behavior is given with enlargement in one of the corners.

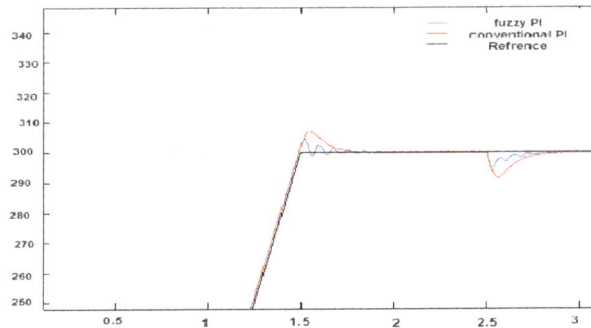

Figure 18. Rotor speed with fuzzy PI controller

As can be seen, the use of fuzzy PI controller significantly improves the system response and the system could follow the reference signal with very good accuracy. By changing the parameters of the membership functions related to the fuzzy inference system, the obtained results can be improved.

References

[1] AR Millner. *Mult hundred horsepower permanent magnet brushless disc motors.* In Proc. IEEE Appl. Power Electron. Conf. (APEC'94). 1994: 351–355.

[2] Kun Wei, Zhengli Lou, Zhongchao Zhang, Research on the Commutation Current Prediction Control in Brushless DC Motor.

[3] N Hemati, MC Leu. A complete model characterization of brushless dc motors. *IEEE Trans. Ind. Applicat.* 1992; 28: 172–180.

[4] P Pillay, R Krishnan. *Modeling, Simulation and Analysis of a Permanent Magnet Brushless dc Motor Drive.* Conference Record of IEEE/IAS Meeting. 1987: 8.

[5] Sung-In Park, Tae-Sung Kim, Sung-Chan Ahn, Dong-Seok Hyun. An Improved Current Control Method for Torque Improvement of High- Speed BLDC Motor.

[6] P Philip, D Meenakshy K. Modelling Of Brushless DC Motor Drive Using Sensored And Sensorless Control. *IEEE Trans. J. Industry Application.* 2012; 2(8).

15

Simulation Analysis of Prototype Filter Bank Multicarrier Cognitive Radio Under Different Performance Parameters

A.S Kang*[1], Renu Vig[2]
[1]ECE Deptt, Panjab University Regional Centre, Hoshiarpur, Punjab, INDIA
[2]ECE Deptt, Panjab University, Chandigarh, INDIA
*Corresponding author, e-mail: askang_85@yahoo.co.in

Abstract

Cognitive Radio has proven as an optimum technique for getting improved spectrum utilization by sharing the radio spectrum with licensed primary users opportunistically. The cognitive radio is a new paradigm to overcome the persisting problem of spectrum underutilization.Seeing the everincreasing demand of wireless applications,the radio sp ectrum is a valuable resource and in cognitive radio systems,trustworthy spectrum sensing techniques are required to avoid any harmful interference to the primary users. As cognitive radio possesses the capability to utilise the unused spectrum holes or white spaces so, there is a tremendous need to scan the large range of spectrum either for interference management or for primary receiver detection. Dynamic Spectrum Access techniques need to be implemented for the sake of better radio resource management and computational complexity analysis of multirate filter bank cognitive radio, where BER and Eb/No are the performance metrics or governing parameters to affect the system performance using polyphase filter bank. The present paper deals with the study of effect of variation of number of subchannels M at fix overlapping factor K of polyphase component of Filter Bank Multicarrier cognitive radio in terms of prototype filter length at Lp=K*M.

Keywords: FBMC, cognitive-radio, spectrum, sub channels, filter length

1. Introduction

The recent interest in cognitive radio based research has attracted a great deal of attention in spectrum sensing and detection of radio users in the environment. The primary objective is to maximize the probability of detection without losing much on the probability of false alarm while minimizing the complexity and time to sense and detect the radio. The limited spectrum for dense wireless communications and inefficient spectrum utilization necessitates a new communication paradigm cognitive radio which can exploit the unutilized spectrum opportunistically. The evolving research efforts are in the field of cognitive radios covering the different Physical Layer Aspects such as cognitive process, different modulation techniques which are utilised for the signal transmission in CR and varying spectrum sensing techniques. CR as a future wireless communication system is characterized as a system,which is able to adapt its transmission and reception parameters on the basis of cognitive interaction with the wireless environment in which it actually operates.There are several reasons that make spectrum sensing, a practically challenging task. A very low required SNR for primary user detection, multi-path fading and time dispersion effects of wireless channels which tend to complicate the spectrum sensing problem and the frequently changing noise level with time and location causes the noise power uncertainty a big issue of concern for primary user detection [1].

2. Physical Layer of Cognitive Radio

Spectrum sensing is the main task of the PHY layer of a cognitive radio. Cognitive radio reflects the situations where both primary and secondary users occupy the same channel space as in licensed band situations. It is also responsible for spectrum sensing and reconfiguration of the transmission parameters. Cognitive Radio can reconfigure its operating frequency, modulation, channel coding and output power without hardware replacement, this is the most significant difference between cognitive radio network and other wireless networks physical layer [2-4]. Software defined radio (SDR) based RF front-end transmitters and receivers are required for configurability of cognitive radio networks. Implementing RF front-end, heavy-weight signal processing algorithms, detecting weak signals, presence of PU while there are secondary users, are significant sensing problems in Cognitive Radio [1].

2.1. Main Features of Physical Layer

PHY Transport: 802.22 use Orthogonal Frequency Division Multiplexing (OFDM) as transport mechanism. Modulation: QPSK, 16-QAM and 64-QAM are supported. Coding: Convolutional Code is Mandatory. Turbo, LDPC or Shortened Block Turbo Codes are optional but recommended. Pilot Pattern: Each OFDM / OFDMA symbol is divided into sub-channels of 28 sub-carriers of which 4 are pilots, which are inserted every 7 sub-carriers. No frequency domain interpolation is required. Net Spectral Efficiency: 0.624 bits/s/Hz –3.12 bits/s/Hz. Spectral Mask: 802.22 have adopted the Spectral Mask requirements proposed by FCC.(200 tap FIR filter required). Figure 1 shows the Block diagram of CR-OSI Model showing the importance of Physical layer.

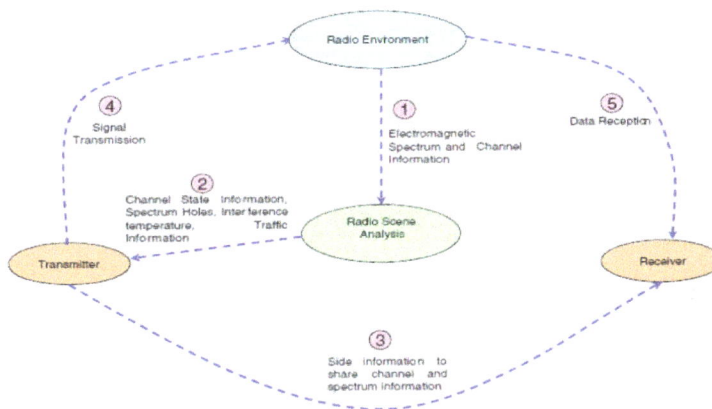

Figure 1. CR Block Diagram of OSI Model

3. Earlier Related Work

Sheryl Ball et al [2005] has focussed on the potential of CR in military and emergency service uses.Some of the issues involved in adapting cognitive radio technology consumer markets were examined and some potential advantages to CR defined networks for consumers were highlighted with some possible service designs and pricing systems [2]. Tero Ihalainen et al [2006] introduced a new low complexity per-subcarrier channel equalizer for FBMC transceiver for high-rate widebandcommunication over doubly-dispersive channel and analyzed its performance. It was shown that the coded error-rate performance of FBMC is somewhat better than that of the OFDM reference [3]. Qiwei Zhang et al [2007] discussed that benefits of CR are clear when used in emergency situations. The idea of applying CR to emergency network is to alleviate this spectrum shortage problem by dynamically accessing free spectrum resources. CR is able to work in different frequency bands and various wireless channels and supports multimedia services such as voice, data and video. Reconfigurable radio architecture is proposed to enable the evolution from the traditional Software Defined Radio (SDR) to CR [4]. Shuang Liu [2008] investigated the motive, definition and paradigms of cognitive radio. The author described the system model used comprising of one primary user and one secondary user with perfect channel side information available at both its transmitter and receiver and channel fading gains [5]. Xin Kang et al [2008] studied the optimal power control policies for fading channels in cognitive radio networks considering both the transmit and interference power constraints. For each of the constraints peak power and average power are investigated. The author derived the optimal power allocation strategies in terms of maximizing the ergodic capacity of the secondary user when channel state information is available to the transmitter and the receiver [6]. Musbah Shaat et al [2009] addressed the problem of resource allocation in multicarrier based CR networks.The objective is to maximize the downlink capacity of the network under both total power and interference introduced to the primary users constraints.The optimal solution has high computational complexity, which makes it unsuitable for practical applications and hence a low complexity suboptimal solution is proposed. The performance of using FBMC instead of OFDM in CR systems is investigated.Simulation results show that the proposed resource allocation algorithm with low computational complexity achieves near optimal performance and proves the efficiency of using FBMC in CR context [7]. Qihang Peng et al [2010] discussed the analysis and simulation of sensing deception in fading cognitive radio networks [8]. M Bahadir Celebi et al [2010] proposed cognitive radio implementation by using standard wireless communication laboratory equipments such as signal generator and spectrum analyzer. Equipment's are controlled through MATLAB instrument control toolbox to carry out CR capabilities specified by IEEE 802.22 WRAN standard. The aim of the work is to provide a CR environment for spectrum sensing algorithms to perform a comparative study considering wireless microphone signals for research and educational purposes [9]. Shixian Wang et al [2011] discussed the cognitive radio simulation environment realization based on autonomic communication [10]. Ajay Kr Sharma et al [2011] analysed the BER performance of Cognitive Radio Physical Layer over Rayleigh channel under different channel encoding schemes,digital modulation schemes and channel conditions. It has been anticipated from simulation study that the performance of the communication system degrades with the increase of noise power [11]. Musbah Shaat and Fauzi Bader [2012] considered a multicarrier based CR network. The goal is to maximize the total sum rate of CR system while ensuring that no excessive interference is induced to the primary system [12]. Satwant Kaur [2013] studied the wireless networks with cognitive radio technology and discussed its various functions and capabilities as a new networking paradigm for future wireless communication [13].

4. Problem Formulation

Large parts of assigned spectrum are underutilized while the increasing number of wireless multimedia applications led to spectrum scarcity. Cognitive radio is an option to utilize non used parts of the spectrum that actually are assigned to primary services. The benefits of cognitive radio are clear in the emergency situations. Current emergency services rely much on the public networks. This is not reliable in public networks where the public networks get overloaded. The major limitation of emergency network needs a lot of radio resources. The idea of applying Cognitive Radio to the emergency network is to alleviate this spectrum shortage problem by dynamically accessing free spectrum resources. Cognitive Radio is able to work in different frequency bands and various wireless channels and supports multimedia services such as voice, data and video. The literature survey shows that the performance of FBMC based CR system can be enhanced under different radio environment. There is a need to work out on the gaps that have been affecting the performance of cognitive radio.The spectral efficiency of OFDM based Cognitive Radio is less as compared to FBMC based CR due to the insertion of Cyclic Prefix in OFDM results in high spectral leakage in its prototype filter. FBMC is severely affected by non linear distortions but the resulting out of channel leakage is still lower compared to OFDM. Hence, non linear distortion must be minimized for efficient FBMC systems. The problem of vertical and horizontal sharing of radio spectrum and transmission power control exists.The problem of Interference in radio environment under different nodes for primary and secondary user locations result in spectrum underutilization.Optimization and tradeoff of power, capacity constraints variables under CR radio environment should be considered.The Resource Allocation Strategies to combat inherent interference on primary user are less efficient.The performance of CR in terms of Energy Spectrogram and Power Spectrogram under different modulation and coding techniques in different radio environment (fading channels) without picocell, microcell, femtocell approach with different indoor and outdoor propagation models using spectrum sensing approach is not effective. Lack of Interlayer Optimization (Cross Layer design) results in degradation in radio resource management. The problem of Congestion in ISM bands adversely affecting the quality of communication persist in Conventional OFDM based network without Dynamic Spectrum Access which is not there in CR based network (with dynamic spectrum access).The present study puts its focus on Performance Enhancement of Filter Bank Multicarrier (FBMC) based Cognitive Radio (CR) in adaptive domain under different strategic conditions of wireless environment. By introducing techniques to improve the spectral efficiency and minimizing the spectrum underutilization, hence improving the overall performance of FBMC based CR.

5. Flowchart for the Present Study

A flowchart has been prepared for study of effect of sub channels on prototype filter length for fbmc cognitive radio. The following system parameters have been used in the present investigation. K-Overlapping Factor, M-Number of Subchannels, D=Channel Delay, Lp=Prototype Filter Length. The distinctive feature of the fbmc design technique has ability to provide improved frequency selectivity through the use of longer and spectrally well shaped prototype filters. In the present case, more emphasis has been laid on the Lp=KM-D as a specific prototype filter length, under the assumption of Delay D as zero.

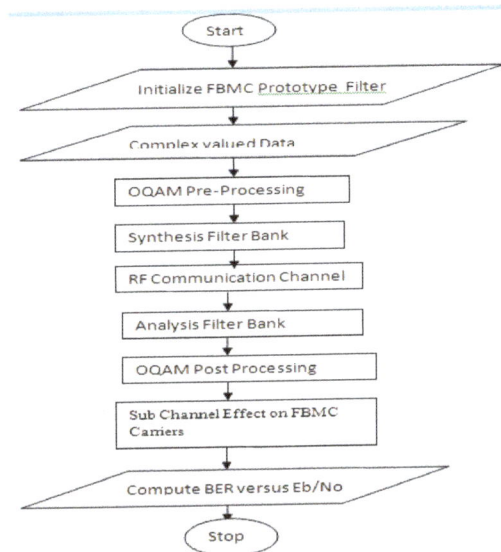

Figure 2. Flowchart for the process flow

5.1. Algorithm for System Level Simulation

```
%Step1 : Define the parameters for OQAM
nd=2; %number of OQAM sample per symbols
M=512;% number of sub-channels
Sub_channels=0:2:M; % Sub_channels
K=4; % overlapping factor(Optimun value at K=4)
Overlapping_factor=K;% Overlapping_factor
lp=K*M-1;% prototype filter length
SNR= 0:5; % signal to noise ratio
nloop=10;% Number of simulation loops
Rate0= zeros(1, length(SNR));
%%Step2:Design of Prototype Filter
%%low pass filter FIR
y=[1 0.971195983 sqrt(2)/2 0.213514695 ]; % coefficient of filter when K=4
yy=[1 0.9111438 0.4111438];% coefficient of filter when K=3
yyy=[1 sqrt(2)/2];% coefficient of filter when K=2
u=lp-1;
s=2*pi/(K*M);
for m=0:u-1
    r=m+1;
```

```
p(m+1)=y(1,1)-2*y(1,2)*cos(r*s)+2*y(1,3)*cos(2*r*s)-2*y(1,4)*cos(3*r*s);% prototype
filter equation K=4
p2(m+1)=yy(1,1)-2*yy(1,2)*cos(r*s)+2*yy(1,3)*cos(2*r*s); % prototype filter equation
K=3
p3(m+1)=yyy(1,1)-2*yyy(1,2)*cos(r*s); % prototype filter equation K=2
end
for m=0:M-1
p4(m+1)=1;
end
%%Step3:Frequency response of prototype filter for FBMC and OFDM
figure,plot(p,'r-^')
hold on
plot(p2,'m-d');
hold on
plot(p3,'b-+');
legend('K=4','K=3','K=2')
hold off
title('Frequency response of prototype filter');
%p=[0 p 0]; % extra delay sample (z^-2)
ty=p;
%Transmitter
%%for ebn0=1:length(SNR)
ebn0
for iii=1:nloop
% Data generation
q=2;% binary data
m=4;% modulation level
In_Data=randi([0 q-1],M,1);% random input data M x 1
    %Step3.1: OQAM Modulation
[OQAM_In_Data]=OQAM_preprocessing(In_Data,m,M); % OQAM modulation M x nd
%%% OQAM_preprocessing
%Transform Block
%% beta carrier
for k=0:M-1
    B(k+1)=(-1)^k*exp(-1*i*2*pi*k*(lp-1)/(2*M));
end
B=reshape(B,M,1);
% ch1=Bi*Di   beta*data
for k=0:M-1
 %   CHT1(k+1,:)=B(k+1)*Train_Data(k+1,:);
    CHD1(k+1,:)=B(k+1)*OQAM_In_Data(k+1,:);
end
%% 'ifft'
    CHD2=ifft(CHD1);
%Polyphase Filters SFB
    for k=0:M-1
    j(k+1,:)=p(k+1+M*(0:K-1));%% Aq coefficients
    ss=upsample(j(k+1,:),2);
    CHD3(k+1,:)=cconv(CHD2(k+1,:),ss,nd);%% ch2*Aq
    end
%P/S Conversion
%%Up sample and delay chain
 % instead of up sample and delay chain
[a b]=size(CHD3);
  CHD4=reshape(CHD3,a*b,1);
 %% channel effects
 CHD5=reshape(CHD4,a*b,1);
 CHD6=CHD5;
```

```
  % CHD7 = awgn(CHD6,ebn0,'measured');
   signal_in_dB=10*log10(std(CHD6)^2);
   noise_in_dB=signal_in_dB-SNR(ebn0);
   noise=(10^(noise_in_dB/10)^(1/2))*randn(size( CHD6, 1),size( CHD6, 2));
   CHD7=CHD6+noise;
%%Receiver:S/P Conversion
%%CHD7=reshape(CHD7,a,b);
%Polyphase Filters AFB
%%for k=0:M-1
   ee=upsample(j(M-k,:),2);
   CHD8(k+1,:)=cconv(CHD7(k+1,:),ee,nd);
   end
%Transform Block
%'fft'
   CHD9=fft(CHD8);
% Beta demodulator
   for k=0:M-1
   BB(k+1)=(-1)^k*exp(1*i*2*pi*k*(lp-1)/(2*M));
end
BB=reshape(BB,M,1);
for k=0:M-1
   CHD10(k+1,:)=BB(k+1)*CHD9(k+1,:);
end
%Step4:OQAM Demodulation
[Out_Data]=OQAM_postprocessing(reshape(CHD10,M,nd),m,M);
%%BER%%[err0, rate0]= symerr( Out_Data,In_Data);
Rate0(ebn0)= Rate0(ebn0) + rate0;
end
% Average value
 Rate0(ebn0)= Rate0(ebn0)/nloop; %
 end
%figure
plot(SNR,Rate0,'r-s')
xlabel( 'Signal-to-Noise Ratio(SNR) in [dB]')ylabel( 'Bit Error Rate (BER)')
title('BER V/s SNR');hold on;grid
on;complexity_analysis(Sub_channels,Overlapping_factor);
```

6. Results and Discussions

Different Graphic plots have been obtained between the various parameters namely Number of Subchannels M, Bit Error Rate BER, Signal to Noise Ratio SNR, Prototype Filter Length Lp=K*M at a fix value of Overlapping factor K=4. The spectrum sensing is performed by measuring the signal strength at the outputs of the subcarrier channels at the receiver. The Cognitive Radio system is able to transmit over the direct link more than that when the direct link is blocked for all subcarriers in the source side. The impact of the present study of FBMC CR is highlighted through the role of number of subchannels. Readjustment of various parameter levels leads to optimization between different radio environment parameters under varying strategic conditions. The computational complexity of the FBMC cognitive radio is studied under the effect of K, M and Lp. For FBMC system, the prototype coefficients are assumed to be equal to PHYDAS coefficients with overlapping factor K=4. Actually, the entire process here involves the three steps. Fixing the subcarriers, matching the subcarriers and re-adjusting the assigned subcarriers as per the system requirement on an average basis.For optimization and tradeoff sake, the number of subcarriers is taken to be greater than 8. The literature survey on FBMC shows that the different subcarriers are adjusted in such a way that the interference to the primary user by secondary users is kept to a minimum. Moreover,the impact of different constraints values on the system performance is investigated. This paper describes a linear assignment problem to select and match some subcarriers for transmission and use the rest only for direct transmission. The FBMC CR in physical layer is a potential candidate for future wireless communication system. A Bandwidth of 10Mhz with M=64, 128, 256, 512 subcarriers have been taken into consideration in the present scenario. The subcarriers are allocated sequentially to the users with optimum results. Figure 3 shows BER versus SNR plot at different values of M. BER is found to decrease with increase at different values of SNR. The trend for M values has been clearly depicted here, at M=64, BER=1.45 at SNR=0 db, BER=1.1 at SNR=1db, BER=0.75 at SNR=2db, BER=0.45 at SNR= 3db, BER=0.3 at SNR=4db, BER=0.1 at SNR=5db. It is very clear that at higher values of M, initial values of BER are found to be higher than 1.45 at SNR=0db. The values of SNR chosen are well within the range (-5db to +30db) as per specifications of IEEE802.22 Standard for FBMC Cognitive Radio. Figure 4 shows the matlab plot between Lp and BER which shows that beyond Lp=500, BER is found to rise to a value $10^{-0.2}$ till Lp=2000. Figure 5 shows matlab plot between Lp and SNR which clearly shows SNR decrease beyond Lp=10^3. Figure 6 shows matlab plot between M and BER which indicates that beyond M=10^2, BER increases to $10^{-0.2}$. Figure 7 shows with M more than 10^2, Lp becomes more than 10^3. In Figure 8, SNR decreases to 2db beyond M=10^2. Various performance parameters have depicted that with increasing BER, SNR is found to decrease and vice versa.

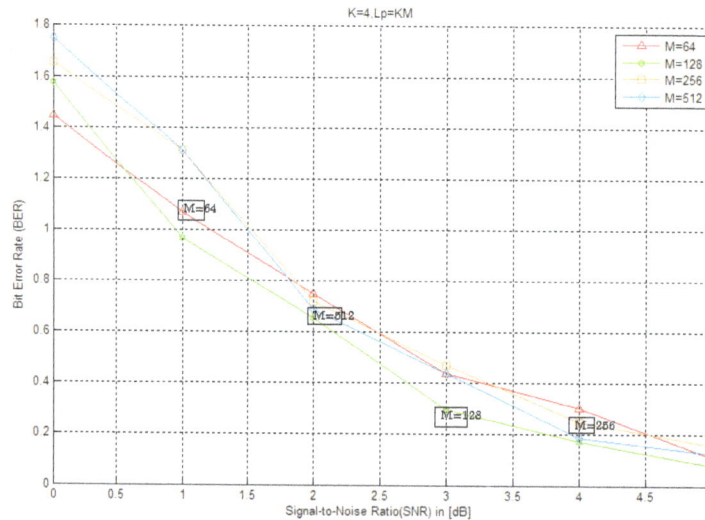

Figure 3. Matlab plot between BER and SNR at K=4, Lp=KM with M=64, 128, 256, 512

Table 1. System Simulation Parameters

Parameter	Value
Total Bandwidth B	5,10 Mhz
Bandwidth per Subcarrier	9-10kHz
Number of Subcarriers/subchannels M	64,128,256,512 (Beyond M=128,Complexity Increases)
Overlapping Factor K	2,3,4,6,8(K=4 Optimum)
Prototype Filter Length Lp=K*M	256,512,1024,2048[At K=4] 192,384,768,1536[At K=3] 384,768,1536,3072[At K=6] 512,1024,2048,4096[At K=8]
Number of OQAM samples per symbol N_d	2,4,5
SNR	-5dB to +30dB
Bit Error Rate	0.45-0.75<1(Desirable)

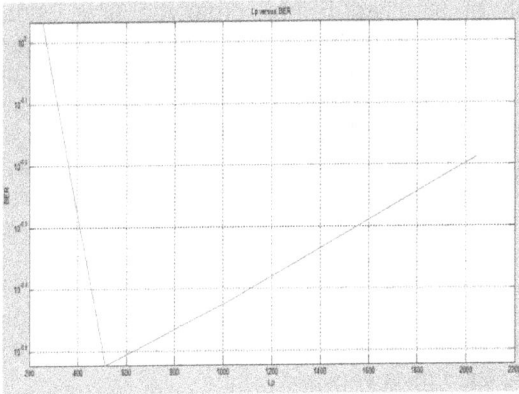

Figure 4. Matlab plot between Lp and BER

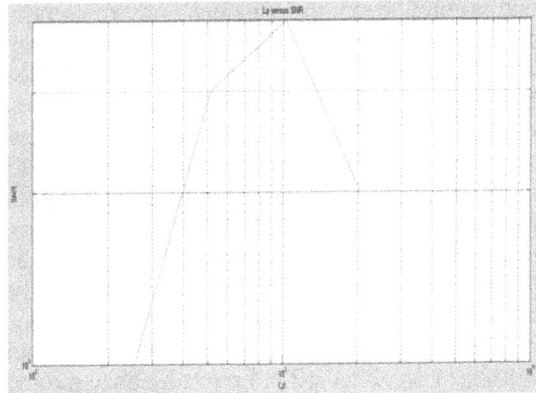

Figure 5. Matlab plot between Lp and SNR

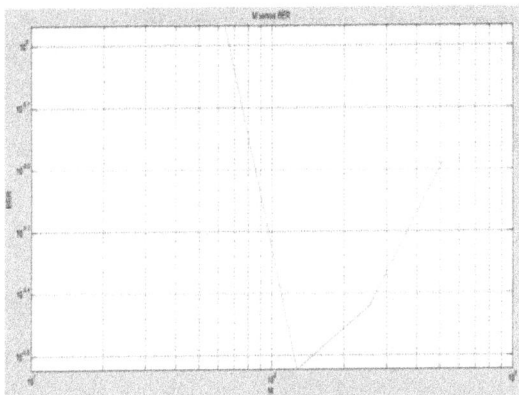

Figure 6. Matlab plot between M and BER

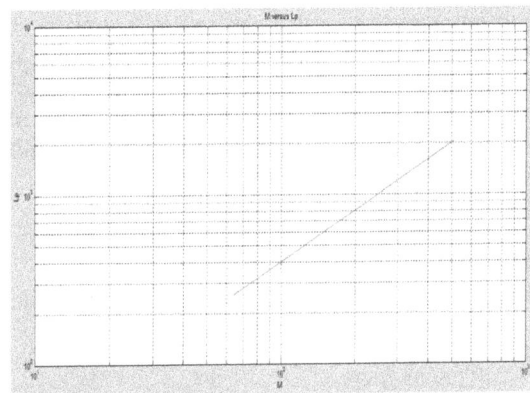

Figure 7. Matlab plot between M and Lp

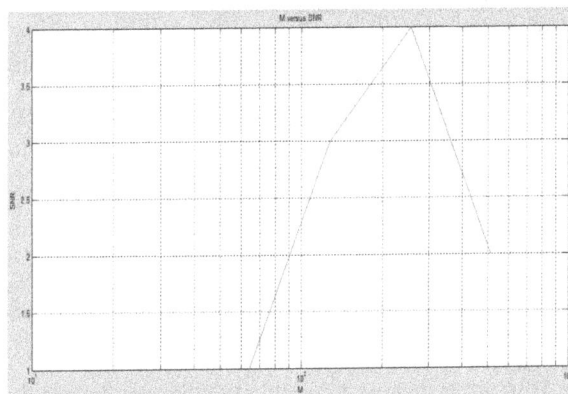

Figure 8. Matlab plot between M and SNR

7. Conclusion

Cognitive radio networks have a promising future and have excellent applications of wireless networks. The signal processing prospectives of CR have significant impact in CR technology in its performance enhancement. The present paper shows the comparatative analysis of number of subchannels M for FBMC prototype filter length at Lp=K*M using BER and Eb/No as performance measuring indicators. Analytical results reveal that the tough requirements on the probability of detection of primary users can only be met with enhancement of channel capacity and spectral efficiency through the variation of Lp, K and M factor corresponding to the design and implementation of modified FBMC prototype filter under different strategic conditions or ubiquitous environment. It is quite apparent that the prototype filter modified version must have the essential characteristics like linear phase with unity roll off to minimize delay and circuit complexity and cost. Moreover, the modified prototype filter should have the transmission zeros at the frequencies which are integral multiples of subchannel spacing, for independent and accurate channel measurements. The main flexibility parameters in the system design are number of subchannels, overlapping factor, which determines the total number of filter coefficients, and the option for partial filter bank design, to equip a low rate user in uplink in an efficient manner. The preliminary results discussed in this paper clearly show that FBMC is sensitive to the RF impairments.The effect of impairments on FBMC cognitive radio can be studied further using analysis and synthesis filter banks with the aid of unified signal models. Comparatative analytical and simulation study of different prototype filter lengths namely, Lp=.KM-1, KM+1 can be easily extended to the future version.

8. Future Directions

As Cognitive radio technology is an important innovation for the future of communications and likely to be a part of the new wireless standards, becoming almost a necessity for situations with large traffic and interoperability concerns. To make Cognitive Radio systems trustworthy, dependable and efficient, a comprehensive energy efficient mechanism is required to identify, remove or mitigate the attacks at any phases of the Cognitive Cycle. The study has its impact on the design and development of cognitive radio system using Filter Bank Multicarrier Approach [14-22].

References

[1] J Mitola, G Maguire. Cognitive Radios: Making Software Radios More Personal. *IEEE Personal. Comm.* 1999; 6(4): 13-18.
[2] Sheryl Ball, et al. *Consumer Applications of Cognitive Radio Defined Networks.* Proceeding IEEE. 2005: 518-525.
[3] Tero Ihalainen, TH Stitz, et al. Channel Equalization in Filter Bank based Multi carrirer Modulation For Wireless Communications. *EURASIP Journal on Advances in Signal Processing.* 2007: 1-18.
[4] Qiwei Zhang, et al. Towards Cognitive Radio for Emergency Networks. *Chapter 1.* 1-26.
[5] Shuang Liu. *Capacity Comparison for Fading Channels under Average and Peak Interference Power Constraints in Underlay Cognitive Radio System.* Proc.Stanford.edu. 2008: 1-8.
[6] Xin Kang, YC Liang, et al. *Optimal Power Allocation for Fading Channels in Cognitive Radio Networks under Transmit and Interference Power Constraints.* Proc. IEEE International Conference on Communications ICC2008. 2008: 3568-3572.
[7] Musbah Shaat, et al. Computationally Efficient Power Allocation Algorithm in Multicarrier- Based Cognitive Radio Networks, OFDM and FBMC systems. *EURASIP Journal on Advances in Signal Processing.* 2010; 2010: 1-13.
[8] Qihang Peng, Pamela C Cosman, et al. *Analysis and Simulation of Sensing Deception in Fading Cognitive Radio Networks.* Proc.2010 IEEE. 2010: 234-242.
[9] M Bahadur Celebi, Huseyin Arslan, et al. *Spectrum Sensing Testbed Design for Cognitive Radio Applications.* Proc. International Conference on Cognitive Radio Applications. 2010: 1-4.
[10] Shixian Wang, Hengzhu Liu, et al. *Cognitive Radio Simulation Environment Realization based on Autonomic Communication.* Proc.2011 IEEE Third International Conference on Communication Software and Networks ICCSN. 2011: 402-407.
[11] Ajay Kr Sharma, et al. BER Performance Analysis of Cognitive Radio Physical Layer over Rayleigh fading channel. *International Journal of Computer Applications.* 2011; 25(11): 25-29.
[12] Musbah Shaat, Fauzi Bader. *Comparison of OFDM and FBMC Performance in Multi-Relay Cognitive Radio Network.* Proc.2012 International Symposium on Wireless Communication Systems. 2012: 756-760.
[13] Satwant Kaur. Intelligent Wireless Network -Cognitive Networks. *IETE Technical Review.* 2013; 30(1): 6-12.
[14] AS Kang, Renu Vig. Analysis of Effect of Variable Number of Subchannels on the Performance of Filter Bank Multicarrier Prototype Filter. *Journal of Electrical & Electronic Systems, Kang and Vig, J Electr Electron Syst.* 2014; 3(1): 1-7.
[15] AS Kang, Renu Vig. BER Performance Analysis of Filter Bank Multicarrier using Sub band Processing for Physical Layer Cognitive Radio. *Journal of Electr Electron Syst.* 2014; 3(3).
[16] AS Kang, Renu Vig. Performance Analysis of Filter Bank Multicarrier Cognitive Radio for Physical Layer under Binary Symmetric Radio Fading Channel. *International Journal of Computer Applications.*

2014; 93(6): 27-32.

[17] AS Kang, Renu Vig. *Computational Complexity Analysis of FBMC-OQAM under Different Strategic Conditions*. Proc.2014 RAECS UIET Panjab University Chandigarh. 2014: 1-9.

[18] AS Kang, Renu Vig. *Study of Filter Bank Multicarrier Cognitive Radio under Wireless Fading Channel*. Proc.IEEE-IACC International Conference. 2014: 209-214.

[19] AS Kang, et al. *Trade-off Between AND and OR Detection Method for Cooperative Sensing in Cognitive Radio*. Proc.IEEE-IACC International Conference. 2014: 395-399.

[20] AS Kang, Jaisukh Paul Singh, et al. Cognitive Radio:State of Research Domain in Next Generation Wireless Networks-A Critical Analysis. *International Journal of Computer Applications*. 2013; 74(10): 1-9.

[21] AS Kang, Rajwinder Singh, et al. Cognitive Radio New Dimension in Wireless Comm-State of Art. *IJCA*. 2013; 74(10): 10-19.

[22] Jaisukh Paul Singh, AS Kang, et al. Cooperative Sensing for Cognitive Radio: A Powerful Access Method for Shadowing Environment, *SPRINGER-Journal of Wireless Personal Communications*. 2014; 2088: 15.

16

Fuzzy Control of Yaw and Roll Angles of a Simulated Helicopter Model Includes Articulated Manipulators

Hossein Sadegh Lafmejani*[1], Hassan Zarabadipour[2]
Faculty of Technical and Engineering, Imam Khomeini International University,
Qazvin, Iran
*Corresponding author, e-mail: h_sadegh@ikiu.ac.ir[1], hassan.zarabadipour@gmail.com[2]

Abstract
Fuzzy logic controller (FLC) is a heuristic method by If-Then Rules which resembles human intelligence and it is a good method for designing Non-linear control systems. In this paper, an arbitrary helicopter model includes articulated manipulators has been simulated with Matlab SimMechanics toolbox. Due to the difficulties of modeling this complex system, a fuzzy controller with simple fuzzy rules has been designed for its yaw and roll angles in order to stabilize the helicopter while it is in the presence of disturbances orits manipulators are moving for a task. Results reveal that a simple FLC can appropriatelycontrol this system.

Keywords: *fuzzy logic controller, helicopter, roll, simmechanics toolbox, yaw*

1. Introduction

Helicopter is a vehicle with high weight and it is designed and made in different types and shapes and for variety of applications like industry, investigation, transportation and many other usages. In other word, helicopter is a multitasking system that is very hard to be controlled in the presence of disturbances and uncertainties. So intelligence control is a very good method for controlling the operation of this system.

There are so many papers about controlling of helicopterslike [1-7], but as I know, there is no paper about controlling of a helicopter includes manipulators in front of it. In order to study a helicopter with this characteristic, it should be simulated and Matlab SimMechanics toolbox is a good software for doing that as it is described briefly in [8].

As we mentioned above, modeling systems like helicopter is troublesome and sometimes impossible using the laws of physics. Therefore, it is not suitable to use classical controllers for these complex nonlinear control applications. Fuzzy logic controller is a suitable, useful and heuristic method for the control of complicated processes in presence of disturbances and uncertainties. In this method there is no need of system modeling or complex mathematical equations governing from the relationship between inputs and outputs. In fuzzy logic we use fuzzy rules easily, even by non- experts. In fuzzy logic the behavior of the system is characterized using human knowledge which leads to the design of control algorithm on the basis of fuzzy rules. As a result, fuzzy logic controller delivers a better performance in cases where the conventional controller does not cope well with non-linearity of a process under control [9]. There are so many papers published using the fuzzy logic controller for controlling the desired systems [1, 5], [10-16]. So, in this paper we used fuzzy logic controller in order to control the yaw and roll angles of an arbitrary helicopter includes manipulators simulated with Matlab SimMechanics toolbox.

2. System Description

The system has been considered in this paper is a helicopter model that have two 2-degree of freedom manipulators in the front of it. This system is shown in Figure 1 from designing in SimMechanics toolbox. These two manipulators are used for a specific task that is not important for us in this paper. The theory behind this system is shown in Figure 2 in a simple way that means: when the angles of the manipulators change, the center of gravity of the helicopter changes too. So we need enough torque for the main and tail rotors to make the helicopter fixed in its position in space.

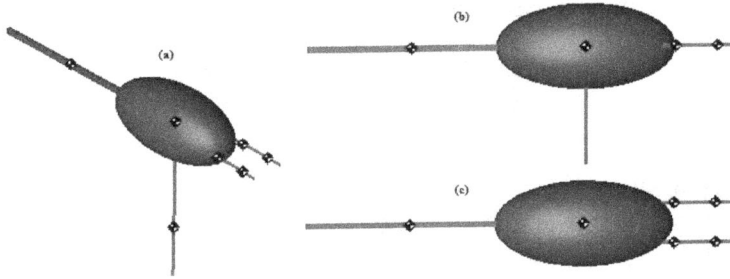

Figure 1. Helicopter model with manipulator, (a) isometric view (b) front view (c) top view

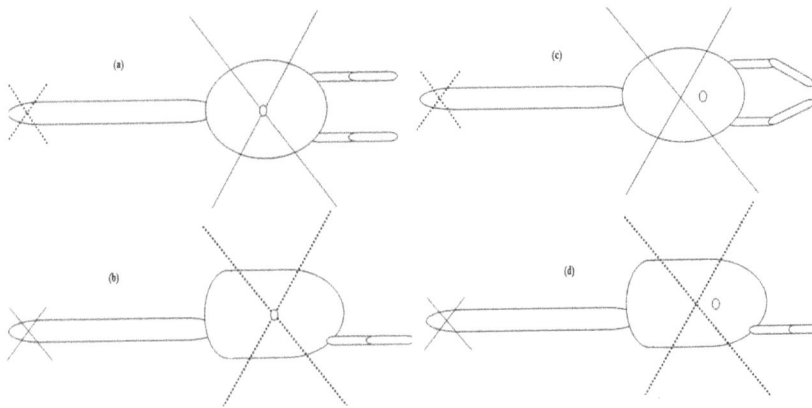

Figure 2. Changes of center of gravity (a) and (b) before manipulator movement, (c) and (d) after manipulator movement

3. Simulation of System with SimMechanics

In order to have a visual view of the system, we need to use a good toolbox to design the system. SimMechanics of Matlab software is a complete toolbox that we can use in order to have the helicopter and its manipulators simulated.

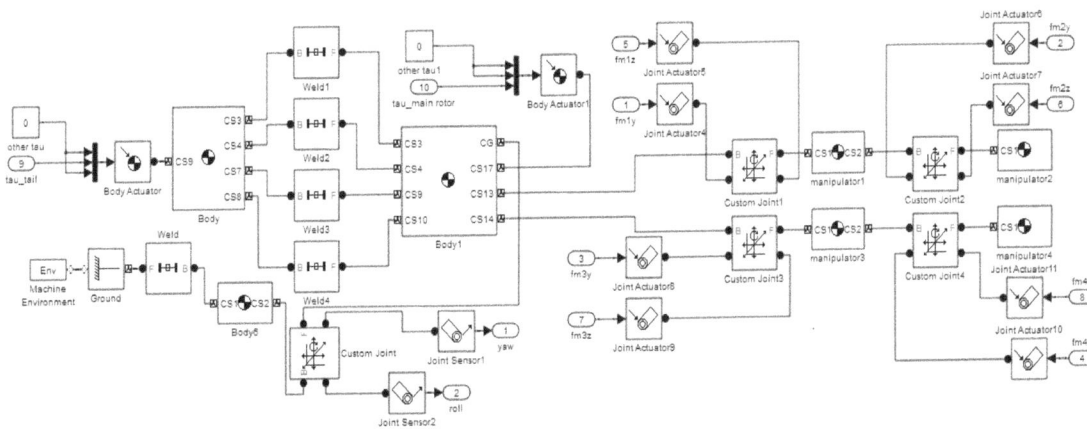

Figure 3. Schematic of the helicopter model and its manipulators in SimMechanics

SimMechanics is based on simulink, which is the research and analysis environment of the controller and the object system in a cross-cutting/interdisciplinary [8]. Multi-body daynamic mechanical systems can be analyze and modeled by SimMechanics and all works such as control would be completed in the simulink envirement. This toolbox provides a plenty number of corresponding real system components, such as: bodies, joints, constraints, coordinate systems, actuators and sensors. Complex mechanical system can be created by these modules in order to analyze them. In this paper, the toolbox has been used to analyze the helicopter model with articulated manipulators. Figure 3 illustrated the schematic of Helicopter model designed in SimMechanics.

4. Fuzzy Logic Controller

The FLC system, first proposed by Zadeh [17]. The system is also known as fuzzy control system or fuzzy inference system or approximate reasoning or expert system. The general framework of this system is shown in Figure 4. According to this framework, we designed the close loop system in Matlab which is illustrated in Figure 5.

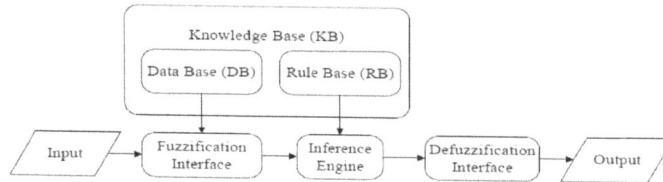

Figure 4. Framework of fuzzy logic controller

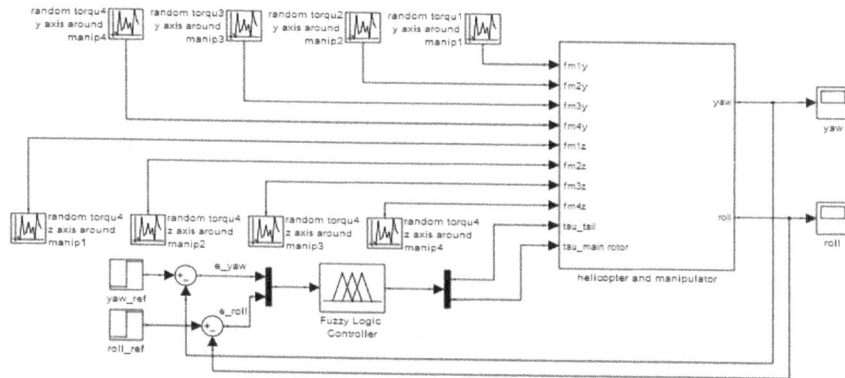

Figure 5. Schematic of fuzzy control system in Matlab Simulink

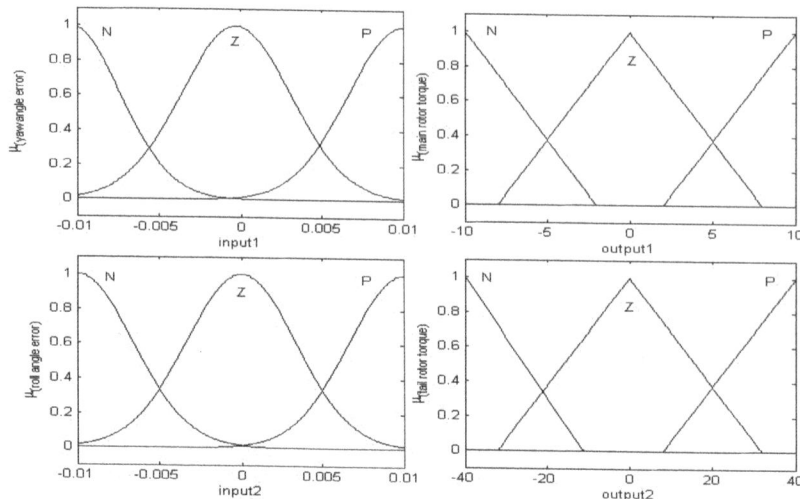

Figure 6. Membership functions of inputs and outputs

In this paper, for the fuzzification process, the Gaussian membership functions and the Triangular membership functions are devoted to the inputs and outputs respectively with the universe of discourse as follows:

$$e = [-0.01, 0.01]deg, \ \tau_{yaw} = [-10, 10]N.m, \ \ \tau_{roll} = [-40, 40]N.m$$

The defuzzification technique used in this study was Center of Gravity approach. Membership functions of inputs and outputs are illustrated in Figure 6.

The simple rule base used in the controller designing is given as follows:

$$if \ (input1 \ is \ N)then \ (output1 \ is \ P)$$
$$if \ (input1 \ is \ P)then \ (output1 \ is \ N)$$
$$if \ (input1 \ is \ Z)then \ (output1 \ is \ Z)$$

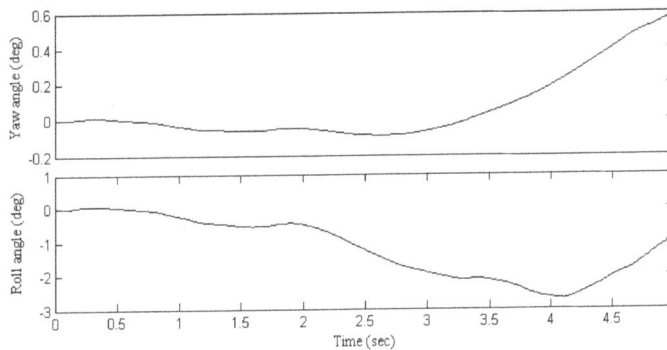

Figure 7. Changes of angles before adding a fuzzy controller

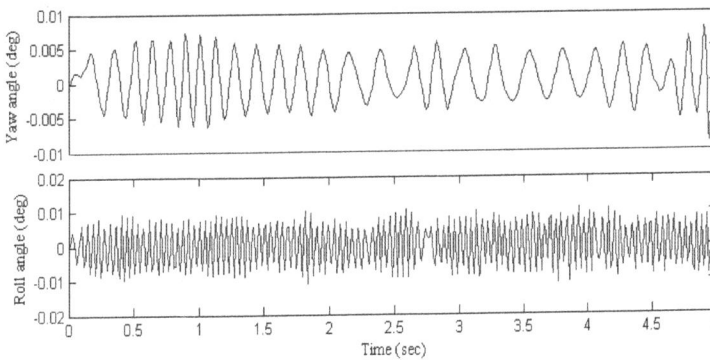

Figure 7. Changes of angles after adding a fuzzy controller

$$if \ (input2 \ is \ N)then \ (output2 \ is \ P)$$
$$if \ (input2 \ is \ P)then \ (output2 \ is \ N)$$
$$if \ (input2 \ is \ Z)then \ (output2 \ is \ Z)$$
$$if \ (input1 \ is \ N)and \ (input2 \ is \ N) \ then \ (output1 \ is \ P) \ and \ (output2 \ is \ P)$$
$$if \ (input1 \ is \ P)and \ (input2 \ is \ P) \ then \ (output1 \ is \ N) \ and \ (output2 \ is \ N)$$
$$if \ (input1 \ is \ Z)and \ (input2 \ is \ Z) \ then \ (output1 \ is \ Z) \ and \ (output2 \ is \ Z)$$

In this paper, our goal was to make the yaw and roll angle of the helicopter at zero(deg) and according to the designed controller we got a good response that is illustrated in Figure 7 and 8. Figure 7 shows the open loop operation of the system and Figure 8 shows the close loop system by adding a fuzzy controller. These results confirm that the fuzzy logic controller is suitable for controlling this system.

5. Conclusion

In this paper, a new mechanical system has been introduced for modeling and applying different kinds of controllers on it. An attempt to control the yaw and roll angles of an arbitrary simulated helicopter includes manipulator by fuzzy logic has been proposed. From the simulation results, it has been shown that the fuzzy logic controller can make a suitable regulation in this system.

References

[1] Phillips C, Karr CL, Walker G. Helicopter flight control with fuzzy logic and genetic algorithms. *Engineering Applications of Artificial Intelligence*. 1996; 9(2): 175-184.

[2] Marconi L, Naldi R. Robust full degree of freedom tracking control of a helicopter. *Automatica*. 2007; 43(11): 1909-1920.

[3] Liu C, Chen WH, Andrews J. Tracking control of small-scale helicopters using explicit nonlinear MPC augmented with disturbance observers. *Control Engineering Practice*. 2012; 20(3): 258-268.

[4] Lee CT, Tsai CC. Nonlinear adaptive aggressive control using recurrent neural networks for a small scale helicopter. *Mechatronics*. 2010; 20(4): 474-484.

[5] Sanchez EN, Becerra HM, Velez CM. Combining fuzzy, PID and regulation control for an autonomous mini-helicopter. *Information Sciences*. 2007; 177(10): 1999-2022.

[6] Hernandez-Gonzalez M, Alanis AY, Hernandez-Vargas EA. Decentralized discrete-time neural control for a Quanser 2-DOF helicopter. *Applied Soft Computing*. 2012; 12(8): 2462-2469.

[7] Furuta K, Ohyama Y, Yamano O. Dynamics of RC helicopter and control. *Mathematics and Computers in Simulation*. 1984; 26(2): 148-159.

[8] Zheng-wen LI, Guo-liang Zhang, Wei-ping Zhang, Bin JIN. *A Simulation Platform Design of Humanoid Robot Based on SimMechanics and VRML*. Procedia Engineering. 2011; (15): 215-219.

[9] Linear Control system Analysis and Design: Conventional and Modern. 4[th]Edition: McGraw-Hill Companies.1995.

[10]]van der Wal AJ. Application of fuzzy logic control in industry. *Fuzzy Sets and Systems*. 1995; 74(1): 33-41.

[11] Bortolet P, Merlet E, Boverie S. Fuzzy modeling and control of an engine air inlet with exhaust gas recirculation. *Control Engineering Practice*. 1999; 7(10): 1269-1277.

[12] Emami MR, Goldenberg AA, Türksen IB. Fuzzy-logic control of dynamic systems: from modeling to design. *Engineering Applications of Artificial Intelligence*. 2000; 13(1): 47-69.

[13] Tunstel E, Hockemeier S, Jamshidi M. Fuzzy control of a hovercraft platform. *Engineering Applications of Artificial Intelligence*. 1994; 7(5): 513-519.

[14] McLean D, Matsuda H. Helicopter station-keeping: comparing LQR, fuzzy-logic and neural-net controllers. *Engineering Applications of Artificial Intelligence*. 1998; 11(3): 411-418.

[15] Babuška R, Verbruggen HB. An overview of fuzzy modeling for control. Control *Engineering Practice*.1996; 4(11): 1593-1606.

[16] Fraichard Th, Garnier Ph. Fuzzy control to drive car-like vehicles. *Robotics and Autonomous Systems*. 2001; 34(1): 1-22.

[17] Zadeh L. Outline of a new approach to the analysis of complex systems and decision processes. IEEE *Trans. Systems, Man Cybernet SMC-3*. 1973; 28-44.

17

Performance of FACTS Devices for Power System Stability

Bhupendra Sehgal*[1], S P Bihari[2], Yogita Kumari[3], R.N.Chaubey[4], Anmol Gupta[5]
[1,2,3]Inderprastha Engineering College, Ghaziabad, India
[4,5]KIET Ghaziabad, India
*Corresponding author, e-mail: bhupendra_sehgal@rediffmail.com[1], spbiharinit@gmail.com[2],
yogitaaryanipec@gmail.com[3]

Abstract

When a power grid is connected to an induction type wind electric generator (WEG), when there is variation in load and wind speed, grid voltage also varies. In this paper, we study what is the impact when there is a variation of load and wind by variation of real power and reactive power consumed by WEG effect of load and wind speed variations on real power supplied and reactive power consumed by the WEG as well as voltage on the grid are studied. The voltage variation in the grid is controlled by reactive power compensation using shunt connected Static VAR Compensator (SVC) comprising Thyristor Controlled Reactor (TCR) and Fixed Capacitor (FC). With the help of Fuzzy Logic Controller (FLC), TCR is operated automatically.

Keywords: *wind electric generator, thyristor controlled reactor, fixed capacitor, fuzzy controller, reactive power*

1. Introduction

In the present power scenario, the demand for electrical power is increasing and conventional resources are depleting fast. Recent studies indicate that there are substantial improvements in the utilization of renewable energy sources especially in the developing countries. This is because of the recent technological developments and also due to environmental and social considerations. Wind energy has experienced remarkable growth over the last decade due to renewed public support and maturing turbine technologies. Many wind turbines use a squirrel cage induction generator to produce electricity. These generators allow small variations in rotor speed thus reducing torque shocks caused by wind gusts [1]. However they absorb large amounts of reactive power and cause severe voltage stability problems on the grid. Traditional shunt compensation using fixed capacitors, due to varying load conditions on the grid and sudden wind gust, sometimes leads to large voltage fluctuations in the grid as the reactive power consumed by the WEG varies widely. So, FACTS devices such as SVC and STATCOM have been suggested as sources for reactive power support due to the ability of these to provide continuously variable susceptance and also the ability to react fast. This paper reports the voltage fluctuations on the grid connected to wind farms due to variations in wind speed and also in loads connected to the grid. The proposed voltage regulation scheme uses a Static VAR Compensator (SVC). SVC comprises a fixed capacitor in parallel with a Thyristor Controlled Reactor (TCR which in turn consists of a reactor in series with a pair of anti parallel thyristors, in each of the three phases.

By varying the firing angle of the thyristor, the fundamental reactive current drawn by the TCR and thereby the net reactive power contributed by the SVC is controlled [2].

2. Wind Electric Generator

A wind turbine is a device that converts kinetic energy from the wind into mechanical energy; a process known as wind power. If the mechanical energy is used to produce electricity, the device may be called a wind generator or wind charger. If the mechanical energy is used to drive machinery, such as for grinding grain or pumping water, the device is called a windmill or wind. Developed for over a millennium, today's wind turbines are manufactured in a range of vertical and horizontal axis types [3]. The smallest turbines are used for applications such as battery charging or auxiliary power on sailing boats; while large grid-connected arrays of turbines are becoming an increasingly large source of commercial electric power.

3. Static Var Compensator (SVC)

A static Var compensator (or SVC) is an electrical device for providin fast-acting reactive power on high-voltage electricity transmission networks. SVCs are part of the Flexible AC transmission system device family, regulating voltage and stabilizing the system. Unlike a synchronous condenser which is a rotating electrical machine, a "static" VAR compensator has no significant moving parts (other than internal switchgear) [4]. Thyristors in anti parallel are used to switch on a capacitor or reactor unit in stepwise control. When the firing angle of thyristors is varied the reactor unit acts as continuously variable in the power circuit. The SVC is an automated impedance matching device, designed to bring the system closer to unity power factor. SVCs are used in two main situations:

a) Connected to the power system, to regulate the transmission voltage ("Transmission SVC")

b) Connected near large industrial loads, to improve power quality ("Industrial SVC")

3.1. Thyristor Controlled Reactor

A thyristor-controlled reactor (TCR) is a reactance, which is connected in series with a bidirectional thyristor valve. The thyristor-controlled reactor is an important component of a Static VAR Compensator.

The thyristor valve is phase-controlled. In parallel with the circuit consisting of the series connection of the reactance and the thyristor valve, there may be an opposite reactance, usually consisting of a permanently connected, mechanically switched or thyristor switched capacitor [5]. By phase-controlled switching of the thyristor valve, the value of delivered reactive power can be set.

3.2. Fixed Capacitor

A capacitor of fixed value is connected in parallel to the network whose value depends upon the total reactive power that has to be supplied. In general instead of a single capacitor, a capacitor bank is employed so that the size of inductor can be smaller, reactive power injected can be regulated smoothly and the amount of ohmic power loss can be reduced.

3.3. Combined Fixed Capacitor – TCR

The fixed capacitor always supplies a constant reactive power (which is equal to the maximum reactive power consumed by the load) to the network. If the reactive power required in the network is lesser than that, the TCR is made to absorb the extra reactive power [6]. This is done by reducing the firing angle of the TCR. If the reactive power required becomes higher, the TCR is made to absorb less which is done by increasing the firing angle.

4. Fuzzy Logic Controller

Fuzzy logic is widely used in machine control. The term itself inspires a certain skepticism, sounding equivalent to "half-baked logic" or "bogus logic", but the "fuzzy" part does not refer to a lack of rigor in the method, rather to the fact that the logic involved can deal with fuzzy concepts—concepts that cannot be expressed as "true" or "false" but rather as "partially true" [7]. A fuzzy control system is a control system based on fuzzylogic—a mathematical system that analyzes analog input values in terms of logical variables that take on continuous values between 0 and 1, in contrast to classical or digital logic, which operates on discrete values of either 1 or 0 (true or false respectively).

5. Development of WEG Model

Cp-λ characteristics of the wind turbine, which is required to model the wind turbine can be obtained from the power curve of a WEG. Figure 1 shows the power curve of the 250kW WEG chosen for the study. Assuming the efficiency of the 250 kW induction generator as 90%, the turbine output is calculated as (250 /0.9) =277.78 kW [8]. The power output is constant from rated wind speed to cut off wind speed. Different values of wind speed in this range are used for calculating the values of Cp and λ.

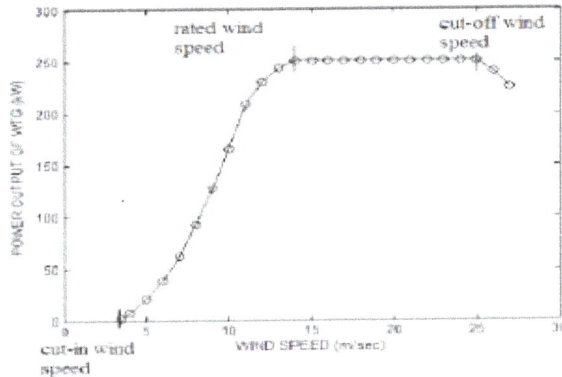

Figure 1. Power curve showing different ranges of wind speed

The Cp-λ graph thus evolved is then extrapolated based on the following assumptions:
a) The maximum Cp of the turbine = 0.45
b) The Nominal λ at which the corresponding Cp is at its maximum = 5
c) The value of λ at which Cp is zero =10.
Figure 2 shows the extrapolated graph. A polynomial expression of Cp in terms of λ is obtained from the graph using curve fitting methods and that is given in Equation (1).

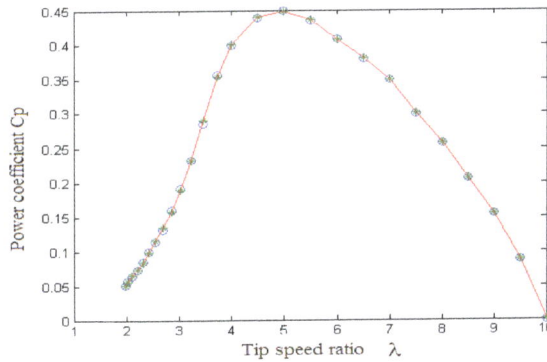

Figure 2. Graph deduced from the power curve

$$Cp = 0.00017\ \lambda 10 - 0.00033\ \lambda 9 - 0.00066\ \lambda 8 + 0.025\ \lambda 7 - 0.044\ \lambda 6 - 0.097\ \lambda 5 \\ + 0.21\ \lambda 4 + 0.088\ \lambda 3 - 0.43\ \lambda 2 + 0.04\ \lambda + 0.4 \qquad (1)$$

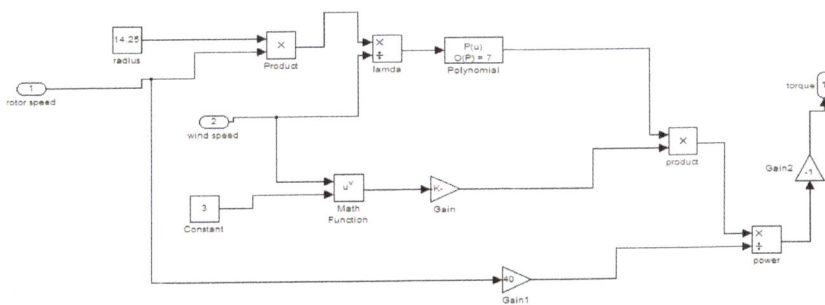

Figure 3. Simulink model of wind turbine

   This expression is used in developing a simulation model of wind turbine. Figure 3 shows the Simulink model of the wind turbine.

6. Simulink Model of Wind Turbine

   Figure 4 shows the system under consideration for the simulation study. The WEG is connected to the power Grid through a transmission line feeding RL load. Figure 5 shows the real power supplied by the grid (P), reactive power absorbed by the WEG (Q) and the grid voltage (V) for different wind speeds without compensation. It is to be noted that the maximum variations in P, Q and V are respectively 0.862 pu, 0.373 pu and 0.021 pu between 7 m/s and 23 m/s. Grid voltage varies from 387V to 378.6V. P and Q vary from 31.95kW to 230.8kW and 61.78kVAR to 98.58kVAR respectively. It is found that the maximum reactive power absorbed by the WEG is 1pu (98.58kVAR) at 14m/s. This is supplied by the reactive power source at the sending end.

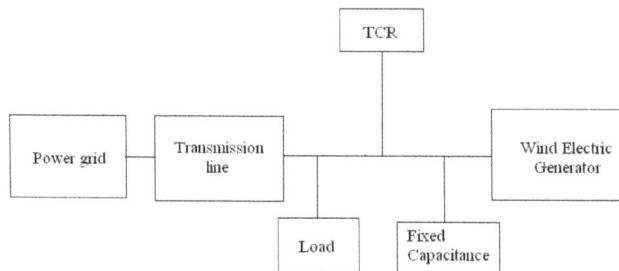

Figure 4. Block Diagram of Wind Electric Generator connected to grid with FC and TCR

7. Block Diagram

   Figure 6 shows the variation in grid voltage due to change in load conditions when the wind speed is kept constant at 14m/sec. It is observed that there is a substantial change of 0.051pu in grid voltage for the load change from 35% to 115%. Grid voltage varies from 414.7V to 394.5V. The results of the study made so far established that FC compensation improves grid voltage substantially, yet it cannot maintain grid voltage constant when there is variation either in wind Speed or in the load demand. Use of TCR along with FC can regulate grid voltage more precisely.

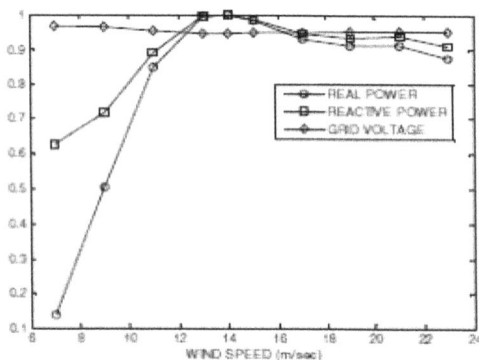

Figure 5. Real power, reactive power for different wind speed

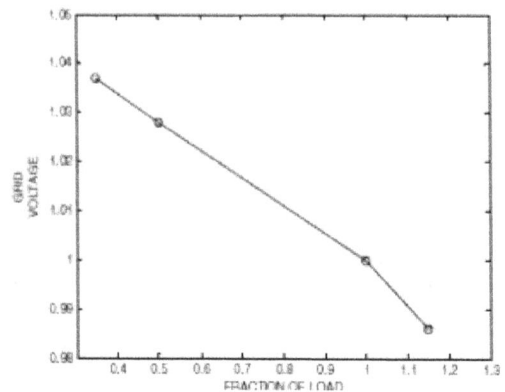

Figure 6. Variation in Grid voltage due to change in load conditions

8. Firing Pulse Generation for TCR

From the above results, it is observed that due to variations in the wind speed and load, the reactive power consumption and therefore the grid voltage varied. For complete and smooth compensation the reactive power supplied should vary as the Q demand. But the reactive power supplied by FC cannot vary. Therefore a three phase star connected Thyristor Controlled Reactor (TCR) is designed and connected at PCC. TCR absorbs the excess reactive power supplied by the Fixed Capacitor (FC). Figure 4 shows the block diagram of WEG connected to Grid with FC and TCR. Whenever there is a change in the Q demand, the firing angle of the TCR has to be varied accordingly in order to maintain the grid voltage constant. To achieve this automatically, Fuzzy Logic Controller (FLC) is implemented. The controller needs to have only one input which is the grid voltage and the single output, which is the firing angle of TCR.

9. Simulation Circuit of the Complete System with Fixed Capacitor

Figure 7 shows the simulation circuit of the complete system with fixed capacitor. In this by varying the capacitor value reactive power, grid voltage & real power of the system is compensated. Figure 8 shows the graph reactive power Vs wind speed, real power Vs wind speed & grid voltage Vs wind speed. As such we change the value of capacitor reactive power of the system change. Figure 8 shows the output of real power, reactive power, & grid voltage i.e. 1.346, 134.6, 211.9. respectively at wind speed 16 m/s. It show that after some fluctuation grid voltage & power in system is constant.

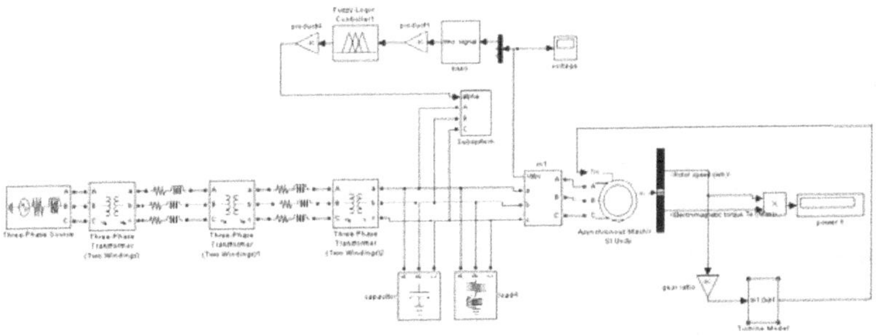

Figure 7. Simulation circuit of the complete system with fixed capacitor

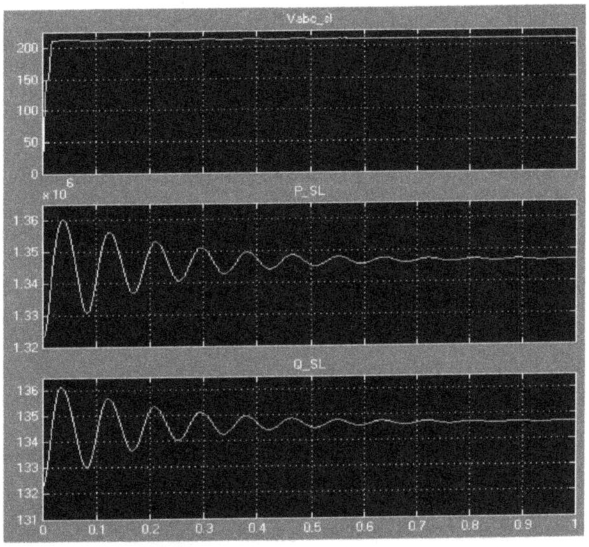

Figure 8. Output of Real Power, Reactive Power, & Grid Voltage

Figure 9 shows the out of real power, reactive power & grid voltage i.e. 1.347 VAR, 134.2, & 212 Volt at wind speed 20 m/s. when the value of capacitor change i.e. increase compensate reactive power is generated in the system.

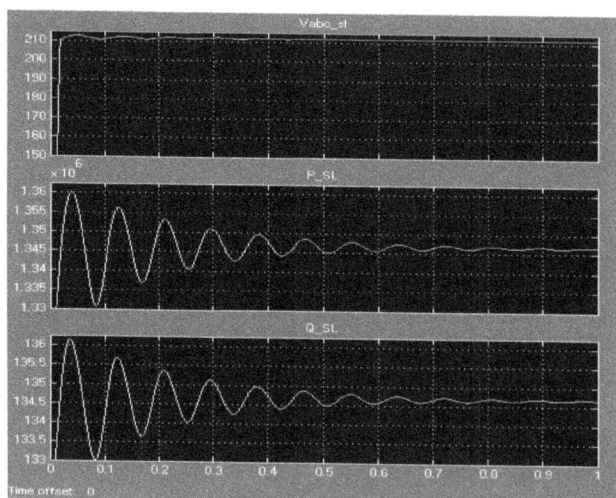

Figure 9. Output of real power, reactive power, grid voltage

10. Conclusion

A methodology to investigate the wind power influences on the power system was presented in this thesis. It includes analysis of the wind power influences on the voltage stability, power system stability and power quality characteristics. The voltage stability was analysed with load ability curves to the power system. The voltage stability was influenced by the wind power integration, where the reactive power was the main factor. A load ability computation tool was developed in this thesis and a static model to the wind power on the voltage stability was presented. Modifications of the wind turbine characteristics, i.e. application of power electronics, were simulated improving the voltage stability. The main conclusions for the voltage stability are that although the wind power alleviates the active power fluxes in the network, the reactive power flux to the wind farms will reduce the voltage stability limits. Wind turbine technologies with power converters that can actively control the reactive power consumption increased the voltage stability (i.e. extended the power limit of the voltage collapse) of the power system. The power system stability and power quality were investigated with dynamic simulations.

References

[1]   Jonathan D Rose, Ian A Hiskens. *Challenges of Integrating Large Amounts of Wind Power*. 1st Annual IEEE systems conference. USA. 2007.
[2]   THE Miller. Reactive Power Control in Electrical systems. John Wiley and Sons. 1982.
[3]   NG Hingorani, Laszlo gyugyi. Understanding FACTS Concepts and Technology of FACTS.
[4]   KJ Saravana kumar, N Saravana Balaji. Simulation studies on Grid connected Wind Electric Generato.
[5]   Luis Fajardo R, Aurelio Medina, Florin Iov. Transient Stability with Grid Connection and Wind Turbine Drive-Train Effects Lui.
[6]   SK Salman, ALJ Teo. Improvement of Fault Clearing Time of Wind Farm Using Reactive Power Compensation.
[7]   Pablo Ledesma, Julio Usaola. Effect of Neglecting Stator Transients in Doubly Fed Induction Generators Models.
[8]   Le Thu Ha, Tapan Kumar Saha. Investigation of Power Loss and Voltage Stability Limits for Large Wind Farm Connections to a Subtransmission Network.
[9]   P Shi, KL Lo. Effect of Wind Farm on the Steady State Stability of a Weak Grid.

18

Enriched Firefly Algorithm for Solving Reactive Power Problem

K. Lenin[1], B. Ravindhranath Reddy[2], M. Surya Kalavathi[3]
[1] Research Scholar, JNTU, Hyderabad 500 085 India
[2] Deputy Executive Engineer, JNTU, Hyderabad 500 085 India
[3]Department of Electrical and Electronics Engineering, JNTU, Hyderabad 500 085, India
email: gklenin@gmail.com

Abstract

In this paper, Enriched Firefly Algorithm (EFA) is planned to solve optimal reactive power dispatch problem. This algorithm is a kind of swarm intelligence algorithm based on the response of a firefly to the light of other fireflies. In this paper, we plan an augmentation on the original firefly algorithm. The proposed algorithm extends the single population FA to the interacting multi-swarms by cooperative Models. The proposed EFA has been tested on standard IEEE 30 bus test system and simulation results show clearly the better performance of the proposed algorithm in reducing the real power loss.

Keyword: firefly algorithm, optimal reactive power, transmission loss

1. Introduction

Various algorithms utilized to solve the reactive power problem. Various numerical methods like the gradient method [1]-[2], Newton method [3] and linear programming [4]-[7] have been utilized to solve the optimal reactive power dispatch problem. The problem of voltage stability and collapse play a   key role in power system planning and operation [8].Evolutionary algorithms such as genetic algorithm have been already utilized to solve the reactive power flow problem [9]-[11]. In [12], Hybrid differential evolution algorithm is utilized to improve the voltage stability index. In [13] Biogeography Based algorithm have been used to solve the reactive power dispatch problem. In [14], a fuzzy based methodology is used to solve the optimal reactive power scheduling method. In [15], an improved evolutionary programming is used to solve the optimal reactive power dispatch problem. In [16], the optimal reactive power flow problem is solved by integrating a genetic algorithm with a nonlinear interior point method. In [17], a pattern algorithm is used to solve ac-dc optimal reactive power flow model with the generator capability limits. In [18], F. Capitanescu proposes a two-step approach to evaluate Reactive power reserves with respect to operating constraints and voltage stability. In [19], a programming based approach is used to solve the optimal reactive power dispatch problem. In [20], A. Kargarian et al present a probabilistic algorithm for optimal reactive power provision in hybrid electricity markets with uncertain loads.This paper proposes Enriched Firefly Algorithm (EFA) to solve reactive power dispatch problem. Our proposed EFA approach is good in exploration and exploitation for searching the global near optimal solution, when compared to other literature surveyed algorithms. A firefly algorithm (FA) is a population-based algorithm enthused by the social behaviour of fireflies [21],[22]. Fireflies converse by flashing their light. Dimmer fireflies are attracted to brighter ones and move towards them to mate [23]. FA is extensively used to solve reliability and redundancy problems. A class of firefly called Lampyride also used pheromone to attract their mate [24]. The proposed Enriched Firefly Algorithm (EFA) extends the single population FA to the interacting multi-swarms by cooperative Models [25]. The proposed EFA algorithm has been evaluated on standard IEEE 30 bus test system. The simulation results show that our proposed approach outperforms all the entitled reported algorithms in minimization of real power loss.

2. Objective function

The Optimal Power Flow problem is considered as a common minimization problem with constraints, and can be written in the following form:

$$\text{Minimize } f(x, u) \tag{1}$$

$$\text{Subject to } g(x,u)=0 \tag{2}$$

And

$$h(x, u) \leq 0 \tag{3}$$

Where f(x,u) is the objective function. g(x.u) and h(x,u) are respectively the set of equality and inequality constraints. x is the vector of state variables, and u is the vector of control variables.

The state variables are the load buses (PQ buses) voltages, angles, the generator reactive powers and the slack active generator power:

$$x = \left(P_{g1}, \theta_2, .., \theta_N, V_{L1}, .., V_{LNL}, Q_{g1}, .., Q_{gng}\right)^T \tag{4}$$

The control variables are the generator bus voltages, the shunt capacitors and the transformers tap-settings:

$$u = \left(V_g, T, Q_c\right)^T \tag{5}$$

or

$$u = \left(V_{g1}, ..., V_{gng}, T_1, .., T_{Nt}, Q_{c1}, .., Q_{cNc}\right)^T \tag{6}$$

Where N_g, N_t and N_c are the number of generators, number of tap transformers and the number of shunt compensators respectively.

2.1. Active power loss

The objective of the reactive power dispatch is to minimize the active power loss in the transmission network, which can be mathematically described as follows:

$$F = PL = \sum_{k \in Nbr} g_k \left(V_i^2 + V_j^2 - 2V_i V_j \cos\theta_{ij}\right) \tag{7}$$

Or

$$F = PL = \sum_{i \in Ng} P_{gi} - P_d = P_{gslack} + \sum_{i \neq slack}^{Ng} P_{gi} - P_d \tag{8}$$

Where g_k: is the conductance of branch between nodes i and j, Nbr: is the total number of transmission lines in power systems. P_d: is the total active power demand, P_{gi}: is the generator active power of unit i, and P_{gsalck}: is the generator active power of slack bus.

2.2. Voltage profile improvement

For minimizing the voltage deviation in PQ buses, the objective function becomes:

$$F = PL + \omega_v \times VD \tag{9}$$

Where ω_v: is a weighting factor of voltage deviation.
VD is the voltage deviation given by:

$$VD = \sum_{i=1}^{Npq} |V_i - 1| \tag{10}$$

2.3. Equality Constraint

The equality constraint g(x,u) of the ORPD problem is represented by the power balance equation, where the total power generation must cover the total power demand and the power losses:

$$P_G = P_D + P_L \tag{11}$$

2.4. Inequality Constraints

The inequality constraints h(x,u) imitate the limits on components in the power system as well as the limits created to ensure system security. Upper and lower bounds on the active power of slack bus, and reactive power of generators:

$$P_{gslack}^{min} \le P_{gslack} \le P_{gslack}^{max} \tag{12}$$

$$Q_{gi}^{min} \le Q_{gi} \le Q_{gi}^{max}, i \in N_g \tag{13}$$

Upper and lower bounds on the bus voltage magnitudes:

$$V_i^{min} \le V_i \le V_i^{max}, i \in N \tag{14}$$

Upper and lower bounds on the transformers tap ratios:

$$T_i^{min} \le T_i \le T_i^{max}, i \in N_T \tag{15}$$

Upper and lower bounds on the compensators reactive powers:

$$Q_c^{min} \le Q_c \le Q_C^{max}, i \in N_C \tag{16}$$

Where N is the total number of buses, N_T is the total number of Transformers; N_c is the total number of shunt reactive compensators.

3. Proposed Enriched Firefly Algorithm

There are three ideal rules amalgamated into the unique Firefly algorithm (FA), i) all fireflies are unisex so that a firefly is attracted to all other fireflies ii) a firefly's attractiveness is proportionate to its brightness seen by other fireflies. For any two fireflies, the dimmer firefly is attracted by the brighter one and moves towards it, but if there are no brighter fireflies nearby means then firefly moves arbitrarily iii) the brightness of a firefly is proportional to the value of its objective function. According to the above three rules, the degree of attractiveness of a firefly is planned by the following equation:

$$\beta = \beta_0 e^{-\gamma r^2} \tag{17}$$

where β is the degree of attractiveness of a firefly at a distance r, β_0 is the degree of attractiveness of the firefly at r= 0, r is the distance between any two fireflies, and γ is a light absorption coefficient.

The distance r between firefly i and firefly j located at x_i and x_j respectively is calculated as a Euclidean distance:

$$r = \|X_i - X_j\| = \sqrt{\sum_{k=1}^{d}(X_i^k - X_j^k)^2} \tag{18}$$

The movement of the dimmer firefly i towards the brighter firefly j in terms of the dimmer one's updated location is determined by the following equation:

$$X_{i+1} = X_i + \beta_0 e^{-\gamma r^2}\left(X_i - X_j\right) + \alpha\left(rand - \tfrac{1}{2}\right) \tag{19}$$

The third term in (19) is included for the case where there is no brighter firefly than the one being considered and rand is a random number in the range of [0, 1].

3.1. Basic Firefly Algorithm

[Step 1] m fireflies are arbitrarily placed within the exploration range, supreme attractiveness is β_0, the light absorption is γ, randomization parameter is α, maximum number of iterations is T; the position of fireflies is arbitrary distributed.

[Step 2] Compute the fluorescence brightness of fireflies. Compute the objective function values of Firefly Algorithm that use the enriched Firefly Algorithm as the largest individual fluorescence brightness value I_o.

[Step 3] Modernize the position of firefly. When the firefly i is no only attracted by a brighter firefly j but also prejudiced by the historical best position of group, the position formula is updating as function (20).The brightest fireflies will modernize their position as the following function:

$$x_{best}(t+1) = x_{best}(t) + \alpha \times (rand - 1/2) \tag{20}$$

Where $x_{best}(t+1)$ is the global optimal position at generation t.

[Step 4] Recalculate the fluorescence brightness value I_o by using the distance measure function $r_{tr}(S)$ after updating the location and penetrating the local area for the toughest fluorescence brightness individual, modernizing the optimal solution when the target value is enriched, or else unchanged.

[Step 5] When reach the maximum iteration number T , record the optimal solution, otherwise repeat step (3), (4), (5) and start the next search. The optimal solution is also the global optimum value H_{max} and the global optimum image threshold is the corresponding threshold value (S, T) at the position $x_{best}(t)$.

3.2. Enriched Firefly Algorithm

In order to overcome the early convergence of classical FA, Cooperative optimization model is amalgamated into FA to construct an Enriched FA in this paper. In order to progress the balance between the exploration and exploitation in EFA we suggest a modification of Eq. (21) used in traditional FA. The goal of the modification is to reinstate balance between exploration and exploitation affording increased possibility of escaping basin of attraction of local optima. In the suggested EFA ,the firefly i find a brighter firefly j when iterative search use Firefly Algorithm ,then move i towards j with a certain step, but the direction of movement will bounce under the influence of the historical best position of group. The direction that i towards j blend with the direction that towards the historical best position of group (x_{best}) is the deflect direction, in this way each search is affected by better solutions thereby improving the convergence rate. The principle of Firefly Algorithm with the influence of the historical best position of group .Suppose any firefly i in the probing range is attracted by a brighter firefly j and influenced by the historical best position of group, then the original direction of movement will change and move towards the optimal direction, thus speeding up the convergence rate.

$$x_i(t+1) = x_i(t) + \beta_0 e^{-\gamma R_j^2} \times \left(x_j(t) - x_i(t)\right) + \beta_1 e^{-\gamma R_{best}^2} \times \left(x_{best}(t) - x_i(t)\right) + \lambda(rand - 1/2) \tag{21}$$

In the proposed EFA, Eq. (20) of old-style FA based on constants values of α and is modified by Eq. (21) Using new variables, λ and β_1 . In this case, the fireflies are adjusted by: Schematic procedure of Firefly Algorithm with the influence of the historical best position of group. The movement of a firefly is attracted to another more brighter firefly with the influence of the historical best position of group is determined by where is the updating position of firefly, $x_i(t)$ is the initial position of firefly which play an important role in balancing the global searching and the local searching, $\beta_o e^{-\gamma R_j^2} \times \left(x_j(t) - x_i(t) \right)$ represents the position of fireflies update under the attraction between fireflies, $\beta_1 e^{-\gamma R_{best}^2} \times \left(x_{best}(t) - x_i(t) \right)$ represent the updating position of fireflies under the influence of the historical best position of group, $\lambda(rand - 1/2)$ is the arbitrary parameter that can avoid the result falling into local optimum.

Input:
Generate preliminary population of fireflies n within
d-dimensional search space x_{ik}, i = 1, 2, . . . , n and
k = 1, 2, . . . , d
Calculate the fitness of the population f (x_{ik}) which
is right proportional to light intensity I_{ik}
Algorithm's parameter— β_o, γ
Output:
Acquired least location: x_i min
start
reiteration
for i = 1 to n
for j = 1 to n
if $\left(I_j < I_i \right)$
Transport firefly i toward j in
d-dimension using Eq. (20)
end if
Attraction varies with distance r
Calculate new solutions and modernize light
intensity using Eq. (21)
end for j
end for i
Rank the fireflies and find the existing best
until stop condition true
end

4. Results and Discussion

EFA algorithm has been verified in IEEE 30-bus, 41 branch system. It has 6 generator-bus voltage magnitudes, 4 transformer-tap settings, and 2 bus shunt reactive compensators. Bus 1 is slack bus and 2, 5, 8, 11 and 13 are taken as PV generator buses and the rest are PQ load buses. Control variables limits are listed in Table 1.

Table 1. Preliminary Variable Limits (PU)

Variables	Min. Value	Max. Value	Type
Generator Bus	0.92	1.12	Continuous
Load Bus	0.94	1.04	Continuous
Transformer-Tap	0.94	1.04	Discrete
Shunt Reactive Compensator	-0.11	0.30	Discrete

The power limits generators buses are represented in Table 2. Generators buses (PV) 2,5,8,11,13 and slack bus is 1.

Table 2. Generators Power Limits

Bus	Pg	Pgmin	Pgmax	Qgmin
1	98.00	51	202	-21
2	81.00	22	81	-21
5	53.00	16	53	-16
8	21.00	11	34	-16
11	21.00	11	29	-11
13	21.00	13	41	-16

Table 3. Values of Control Variables After Optimization

Control Variables	EFA
V1	1.0621
V2	1.0522
V5	1.0317
V8	1.0422
V11	1.0821
V13	1.0641
T4,12	0.00
T6,9	0.02
T6,10	0.90
T28,27	0.91
Q10	0.11
Q24	0.11
Real power loss	4.3001
Voltage deviation	0.9070

Table 3 shows the proposed approach succeeds in keeping the control variables within limits. Table 4 summarizes the results of the optimal solution obtained by various methods.

Table 4. Comparison Results

Methods	Real power loss (MW)
SGA (26)	4.98
PSO (27)	4.9262
LP (28)	5.988
EP (28)	4.963
CGA (28)	4.980
AGA (28)	4.926
CLPSO (28)	4.7208
HSA (29)	4.7624
BB-BC (30)	4.690
EFA	4.3001

5.   Conclusion

In this paper, the EFA has been effectively implemented to solve Optimal Reactive Power Dispatch problem. The proposed algorithm has been tested on the standard IEEE 30 bus system. Simulation results show the robustness of proposed EFA method for providing better optimal solution in decreasing the real power loss. The control variables obtained after the optimization by EFA is within the limits.

References

[1] O. Alsac, B. Scott. Optimal load flow with steady state security. *IEEE Transaction.* 1973; PAS: 745-751.
[2] Lee KY, Paru YM, Oritz JL. A united approach to optimal real and reactive power dispatch. *IEEE Transactions on power Apparatus and systems.* 1985; PAS-104: 1147-1153.
[3] A. Monticelli, MVF. Pereira, S. Granville. Security constrained optimal power flow with post contingency corrective rescheduling. *IEEE Transactions on Power Systems.* 1987; PWRS-2(1): 175-182.
[4] Deeb N, Shahidehpur SM. Linear reactive power optimization in a large power network using the decomposition approach. *IEEE Transactions on power system.* 1990; 5(2): 428-435.
[5] E. Hobson. Network consrained reactive power control using linear programming. *IEEE Transactions on power systems.* 1980; PAS -99(4): 868-877.
[6] KY. Lee, YM. Park, JL. Oritz. Fuel –cost optimization for both real and reactive power dispatches. *IEE Proc.* 131C(3): 85-93.
[7] MK. Mangoli, KY. Lee. Optimal real and reactive power control using linear programming. *Electr.Power Syst.Res.* 1993; 26: 1-10.
[8] CA. Canizares, ACZ. de Souza, VH. Quintana. Comparison of performance indices for detection of proximity to voltage collapse. 1996; 11(3): 1441-1450.
[9] SR. Paranjothi, K. Anburaja. Optimal power flow using refined genetic algorithm. *Electr.Power Compon.Syst.* 2002; 30: 1055-1063.
[10] D. Devaraj, B. Yeganarayana. Genetic algorithm based optimal power flow for security enhancement. IEE proc-Generation.Transmission and. Distribution. 6 November 2005: 152.

[11] A. Berizzi, C. Bovo, M. Merlo, M. Delfanti. A ga approach to compare orpf objective functions including secondary voltage regulation. *Electric Power Systems Research*. 2012; 84(1): 187–194.

[12] CF. Yang, GG. Lai, CH. Lee, CT. Su, GW. Chang. Optimal setting of reactive compensation devices with an improved voltage stability index for voltage stability enhancement. *International Journal of Electrical Power and Energy Systems*. 2012; 37(1): 50–57.

[13] P. Roy, S. Ghoshal, S. Thakur. Optimal var control for improvements in voltage profiles and for real power loss minimization using biogeography based optimization. *International Journal of Electrical Power and Energy Systems*. 2012; 43(1): 830–838.

[14] B. Venkatesh, G. Sadasivam, M. Khan. A new optimal reactive power scheduling method for loss minimization and voltage stability margin maximization using successive multi-objective fuzzy lp technique. *IEEE Transactions on Power Systems*. 2000; 15(2): 844–851.

[15] W. Yan, S. Lu, D. Yu. A novel optimal reactive power dispatch method based on an improved hybrid evolutionary programming technique. *IEEE Transactions on Power Systems*. 2004; 19(2): 913–918.

[16] W. Yan, F. Liu, C. Chung, K. Wong. A hybrid genetic algorithminterior point method for optimal reactive power flow. *IEEE Transactions on Power Systems*. 2006; 21(3): 1163–1169.

[17] J. Yu, W. Yan, W. Li, C. Chung, K. Wong. An unfixed piecewiseoptimal reactive power-flow model and its algorithm for ac-dc systems. *IEEE Transactions on Power Systems*. 2008; 23(1): 170–176.

[18] F. Capitanescu. Assessing reactive power reserves with respect to operating constraints and voltage stability. *IEEE Transactions on Power Systems*. 2011; 26(4): 2224–2234.

[19] Z. Hu, X. Wang, G. Taylor. Stochastic optimal reactive power dispatch: Formulation and solution method. *International Journal of Electrical Power and Energy Systems*. 2010; 32(6): 615–621.

[20] A. Kargarian, M. Raoofat, M. Mohammadi. Probabilistic reactive power procurement in hybrid electricity markets with uncertain loads. *Electric Power Systems Research*. 2012; 82(1): 68–80.

[21] B. Bhushan, SS. Pillai. Particle swarm optimization and firefly algorithm: performance analysis. Proceedings of the 3rd IEEE International Advance Computing Conference (IACC). 2013: 746–751.

[22] PR. Srivatsava, B. Mallikarjun, XS. Yang. Optimal test sequence generation using firefly algorithm. *Swarm and Evolutionary Computation*. 2013; 8: 44-53.

[23] AH. Gandomi, XS. Yang, S. Talatahari, AH. Alavi. Firefly algorithm with chaos. *Communications in Nonlinear Science and Numerical Simulation*. 2013; 18(1): 89-98.

[24] R. De Cock, E. Matthysen. Sexual communication by pheromones in a firefly, Phosphaenus hemipterus (Coleoptera: Lampyridae). *Animal Behaviour*. 2005; 70(4): 807-818.

[25] B. Liu, YQ. Zhou. A Hybrid Clustering Algorithm Based on Firefly Algorithm. *Journal of Computer Engineering and Applications*. 2008; 44(18).

19

Pixel Classification of SAR Ice Images using ANFIS-PSO Classifier

G.Vasumathi*, P.Subashini
Department of Computer Science, Avinashilingam University, Coimbatore, India
e-mail: vasumathidevi@gmail.com

Abstract

Synthetic Aperture Radar (SAR) is playing a vital role in taking extremely high resolution radar images. It is greatly used to monitor the ice covered ocean regions. Sea monitoring is important for various purposes which includes global climate systems and ship navigation. Classification on the ice infested area gives important features which will be further useful for various monitoring process around the ice regions. Main objective of this paper is to classify the SAR ice image that helps in identifying the regions around the ice infested areas. In this paper three stages are considered in classification of SAR ice images. It starts with preprocessing in which the speckled SAR ice images are denoised using various speckle removal filters; comparison is made on all these filters to find the best filter in speckle removal. Second stage includes segmentation in which different regions are segmented using K-means and watershed segmentation algorithms; comparison is made between these two algorithms to find the best in segmenting SAR ice images. The last stage includes pixel based classification which identifies and classifies the segmented regions using various supervised learning classifiers. The algorithms includes Back propagation neural networks (BPN), Fuzzy Classifier, Adaptive Neuro Fuzzy Inference Classifier (ANFIS) classifier and proposed ANFIS with Particle Swarm Optimization (PSO) classifier; comparison is made on all these classifiers to propose which classifier is best suitable for classifying the SAR ice image. Various evaluation metrics are performed separately at all these three stages

Keywords: SAR, speckle noise, particle swarm optimization ice images

1. Introduction

SAR processing is widely used for various radar imagery applications. One of its main applications is monitoring sea ice. Sea ice monitoring has been the main mission objectives for satellite programs such as RADARSAT, European Remote Sensing satellite (ERS) and ENVISAT [1]. Since sea ice has an important role in the heat exchange between the ocean and the atmosphere, it has an impact on the water temperature, heat circulation and ecosystem in Polar Regions. Sea ice is a threat to the navigation and oil exploration, because it can make damage to ships and oil platforms in the sea [2]. Sea ice monitor is useful to know the regions around the ice infested area. It covered with many regions which include sea, vegetation, land etc. For route planning in ship navigation classification of sea ice is an important one.

In this paper pixel based classification is performed where three stages are considered in classifying SAR ice images. It includes Preprocessing, Segmentation and Classification. SAR images are affected by speckle noise thus preprocess stage attempts denoising of speckle noise. Various speckle filters from image processing and wavelet family types are compared in which Daubechies method in wavelet family gives better results in removing noise from SAR ice images. Segmentation process is done by comparing the K-means and Watershed algorithm in which K-means clustering algorithm gives best result in segmenting SAR ice images into various regions. Finally at the classification stage, pixel based classification on the SAR ice images is performed using various supervised learning classifiers such as Back Propagation Neural Network, Fuzzy Classifier, Adaptive Neuro Fuzzy Inference Classifier (ANFIS), and ANFIS with PSO (Particle Swarm Optimization). Experiments at various stages are conducted to obtain results.

Table 1 show the origin of the dataset collected for the SAR ice classification.

Table 1. Origin of the Dataset

No.of Images	Image Location	Satellite	Government
1	Strait of Georgia	SEASAT	Canadian Space Agency, Government of Canada
2	Ward Hunt ice Shelf	RADARSAT-1	Canadian Space Agency, Government of Canada
3	Mouth of the Columbia River and the Oregon coastline	ASF(Alaska Satellite Facility)	NASA, Government of USA
4	Isla Cedros, Baja California	ASF(Alaska Satellite Facility)	NASA, Government of USA
5	Grays Harbor and Willapa Bay in Washington State	ASF(Alaska Satellite Facility)	NASA, Government of USA
6	Early winter sea ice in the Beaufort Sea	RADARSAT-1	Canadian Space Agency, Government of Canada

2. Methodology

Methodology of SAR ice classification includes three stages. Figure 1 shows the overview of the methodology.

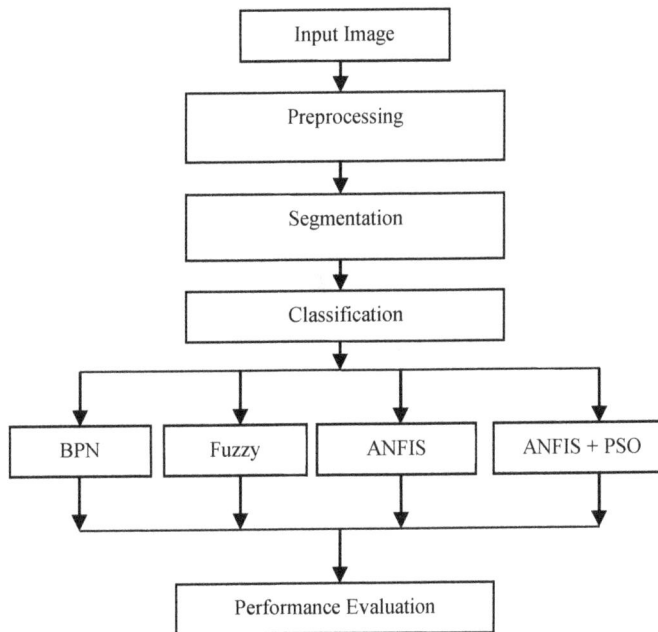

Figure 1. Overview of Methodology

2.1. Preprocessing

SAR images are affected by speckle noise due to the coherent nature of processing the signal [3]. Speckle noise is a coarse noise that is usually evident in and degrades the quality of the active radar and synthetic aperture radar (SAR) images. Speckle noise is commonly observed in radar sensing system and images, although it can be observed in most types of remotely sensed images utilizing coherent radiation. Speckle noise in radar data or images has multiplicative error and must be removed before the data can be used otherwise the noise is merged into and degrades the image quality [4].

In this paper the filters considered are standard speckle removing filters and filter method from wavelet family. This paper involves a comparative method which includes the families in wavelet filter. Through the analysis of this comparison, Daubechies (dB) in wavelet family have proved the best in denoising the image when compared to all other filters. All the filter algorithms are evaluated using software tools called ENVI 4.7 and IDL 7.0. Table 2 shows the various filters taken for removing speckle noise in SAR ice images.

Table 2. Speckle Removal Filters

S. NO	Image processing Filters for speckle control
1	Median Filter
2	Lee Filter
3	Enhanced Lee Filter
4	Frost Filter
5	Enhanced Frost Filter
6	Gamma Filter
7	Kaun Filter
8	Wavelet Family
	Symlet
	Coiflet
	Haar
	Daubechies

The objective results for the preprocessing are given in Table 3.

Table 3. Objective Resluts for Speckle Noise Filter Comparison

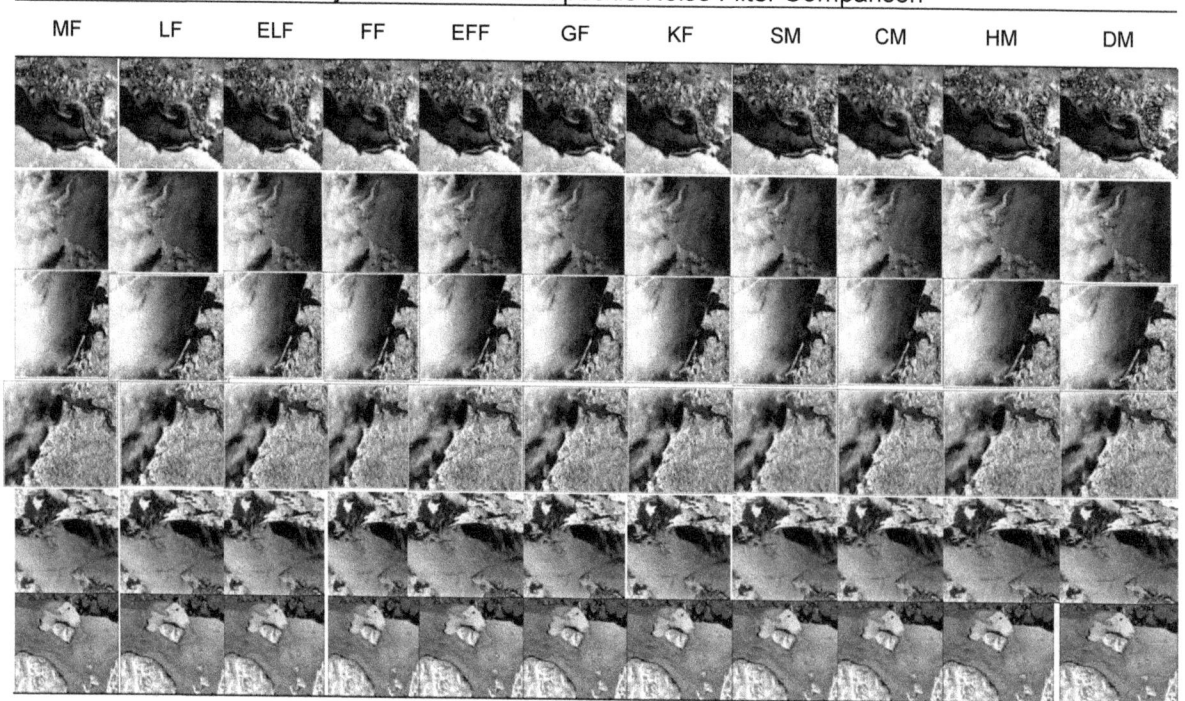

MF- Median Filter, LF- Lee Filter, ELF- Enhanced Lee Filter. GF- Gamma Filter, KF- Kaun Filter, SM-Symlet Method, CM- Coiflet Method, HM- Haar Method, DM- Daubechies Method.

2.2. Segmentation

Image segmentation plays a very important role in the interpretation and understanding of SAR images. It has received an increasing amount of attention over the last few decades [5]. The segmentation procedure is to find the better positions of the shape points according to the appearance information [6]. The purpose of image segmentation is to cluster pixels of an image into image regions.

To classify the SAR ice image it is first needed to segment the regions, further these segmented regions are then identified in the stage of classification. In this paper segmentation of SAR ice images are done using two algorithms namely K-means algorithm and Watershed algorithm. These two algorithms are most promising in segmenting remote sensing images. Comparison of these two algorithms is performed and the result shows that K-means are more efficient in segmenting the images than watershed algorithm. Some of the drawbacks which are find out in Watershed algorithms during the experiment are:

a. It produces over segmentation.
b. It is highly sensitive to the local minima
c. Considerable effort for defining the algorithms takes place for every marking of the watershed.
d. It exhibits the neck line on every region to estimate the number of objects in the given cluster. Errors are produced while exhibits the line on regions.

K-means clustering concept gives more advantages in grouping the pixels of an image to the corresponding classes. In this paper, for segmentation comparison is made in terms of segmentation metrics such as True Positive area ratio, False Positive area ratio and Similarity Index area ratio. The Objective result for comparative analysis between these two segmentation algorithms is given in the following Table 4.

Table 4. Objective Results for Comparison of K-Means

| K-Means Segmented Image | Watershed Segmented Image |

2.3. Classification

Classification process is the main phase in performing SAR ice classification. Classification helps in finding out the features of the ice images such as water, vegetation, land and ice. In classification the segmented regions are identified and predicted. It groups the features based on supervised learning. This process automatically creates a classification model from a set of records called a training set. Once a model is induced, it can be used automatically to classify records belonging to a small set of class that is predefined called a testing set. Training refers to building a new model by using historical data and testing refers to trying out the model on new, previously unseen data to determine its accuracy [7].

In this paper four supervised learning algorithms are performed in the classification of SAR ice images. The comparative analysis between these four classifiers are performed to predict proposed classifier of ANFIS with PSO which gives best classification result in terms of Accuracy, Validity Index (VI), Network sensitivity, Time complexity and Error complexity. Following section gives brief explanation about the classifiers which are performed in this paper.

2.3.1. Back Propagation Neural Network

Back propagation neural classifier is the most simple and best classifier from neural network. It is widely applied for all the classification problems in image processing. The Network Structure of ANN should be simple and easy. There are basically two types of structures recurrent and non-recurrent structure. The Recurrent Structure is also known as Auto associative or Feedback Network and the Non Recurrent Structure is also known as Associative or feed forward Network [8]. Figure 2 shows the Architecture of the Back Propagation Network.

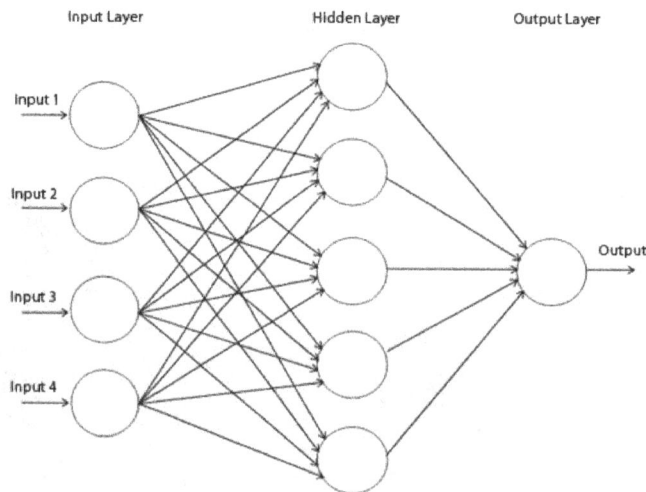

Figure 2. Back Propagation Neural Network Architecture

In this paper, 3-layer Back Propagation network classifier with one output unit is used for the classification purpose. The steps performed in both training and testing process includes,
Step 1: Defining a structure Corresponds to the input and output.
Step 2: Image pixels are given to the network as an input.
Step 3: Assigning the targets according to these inputs.
Step 4: According to the network parameters set, network simulation is performed and an input pixels are trained and store in a file.
Step 5: Target pixels are called and network simulation is presented to the system.
Step 6: During the testing process network simulation is used to compare the test image which is currently selected with the trained pixels.
Step 7: Pixels in a groups are identified if the selected image pixel values are matched with the existing trained pixels.
Step 8: Identified regions are separated and represented with annotations.

During the training process image pixels are given to the network as an input. Targets in which each pixel is stored are assigned. After the training process, classification process is preceded with testing the segmented image with the trained pixel. If training process is correctly done it identifies every region according to the pixel values of an image. The Back Propagation Classifier gives a correct classification rate of 77.14% on the test set images. The experimental results for BPN classifier are given in Table 7 to Table 13. Below are some limitations of BPN classifier, they are,

a. It creates the structure for every incoming input. It does not adjust the incoming inputs with existing structure.
b. Larger number of structure gives complexity in structure. Understanding the concepts will be tough because of its complexity.
c. Errors are increased if the structures are getting increased.
d. Training the pixels takes more time.
e. Parameters need to be assigned to the network to train. Setting the parameter rate is the biggest deal.

2.3.2. Fuzzy Classifier

To overcome the disadvantages in the BPN classifier and to give a highest accuracy in SAR ice classification Fuzzy classifier is considered. Over the past few decades, fuzzy logic has been used in a wide range of problem domains. Although the fuzzy logic is relatively young theory, the areas of applications are very wide: process control, management and decision making, operations research, economies and, for this paper the most important, pattern recognition and classification. The natural description of problems, in linguistic terms, rather than in terms of relationships between precise numerical values is the major advantage of this theory [9]. There are numerous advantages are noticed in fuzzy classifier during the experiment, those are,

a. Fuzzy classifier does not create a new structure for its every incoming input. It adjusts the previous structure according to the incoming inputs.
b. The complexity is very less compare to the BPN.
c. Training taken lesser time compare to BPN.

Fuzzy Inference System diagram is presented in Figure 2.2.

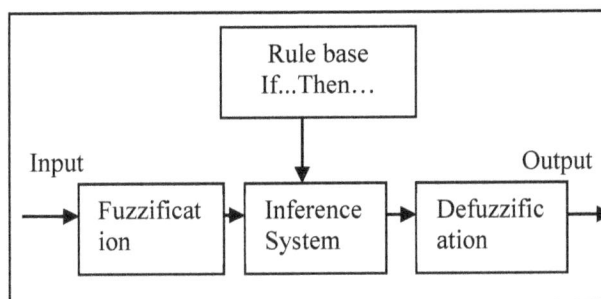

Figure 3. Fuzzy Inferenc System

Fuzzy algorithm steps for SAR ice classification are summarized as follows.
a. Step 1: Defining the input is the first step. Here the inputs are the pixel values.
b. Step 2: The rules with IF, THEN condition are defined in this step. These rules are defined logically to the system in which it can be easily find out the need of the user.
c. Step 3: Third step is defining the membership function that gives the representation to input. Membership function allocated in this paper is the targets in which the pixels should be stored.
d. Step 4: Finally during the testing process original segmented images are compared with the rules which are given to the fuzzy system. At the classification stage features are extracted and represented with annotations.

SAR ice classification using fuzzy classifier makes use of the rules and produces the result accordingly. Training the fuzzy classifier make use of the pixel values of six images. Testing process uses k-means segmented images. Defining a rule gives a better definition for the need of classification. Thus it provides efficient classification which leads to better accuracy than BPN classifier. Fuzzy classifier for SAR ice gives the accuracy rate of 86.6%. Experimental results of fuzzy classifier are given in the Table 7 to Table 13. Fuzzy classifier has limitation and it is listed below.

    a.   Defining the rules in linguistic manner is the only problem in fuzzy classifier.
    b.   Wrong rules can lead to wrong classification.

2.3.3. Adaptive Neuro Fuzzy Inference Classifier (ANFIS)

ANFIS classifier combines both BPN and Fuzzy classifiers advantages. It overcomes the limitations and which gives the better accuracy in classifying SAR ice than other two algorithms discussed above. The combination of fuzzy logic with the design of neural network led to the creation of Neuro- Fuzzy systems which has benefit of the feed forward calculation of the output and back propagation learning capability from the neural networks. Because of crisp consequent functions, ANFIS uses a simple form of scaling implicitly. The ANFIS composes the ability of neural network and fuzzy system [10].

There are some numerous advantages found in ANFIS classifier they are,

    a.   Finding of membership function and appropriate rules in one system can leads to minimize the error.
    b.   BPN can introduce the learning to the fuzzy system which leads to the efficient classification.
    c.   When combining these methods fuzzy system will set a rule for its every input vectors which is then used to calculate the output value.
    d.   Designing a way such that fuzzy system will gets the learning ability of neural network and optimizes its parameters.

2.3.3.1. ANFIS Structure

Designing the ANFIS structure for the SAR ice classification is the initial step to perform the classification. Defining the inputs, rules, membership functions and output for the system is separated according to the five layers. In this paper structure of the ANFIS consists of six inputs and single output. The six inputs represent different pixel values of the images. Each of this training set forms the fuzzy inference system with rules. In fuzzy inference system, fuzzification will take the rules for every input signal. Defuzzification is used to select one particular output signal to the corresponding selected image. Image will be selected during testing process. The basic structure of the ANFIS classifier is presented in the below Figure 4.

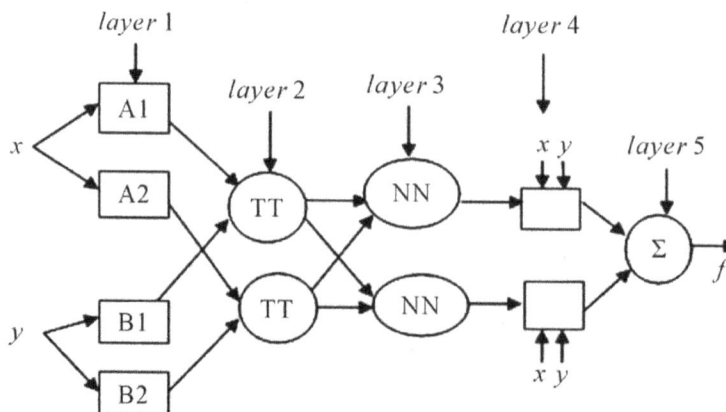

Figure 4. Structure of ANFIS

Summarization of these five layers according to the SAR ice classification includes as follows,

a. Layer 1: In Layer 1 fuzzification process are taken place in which an input of pixels are assigned.
b. Layer 2: In Layer 2 execution of the fuzzy rules are taken place. Rules are assigned according to the inputs and its output.

For ex, if the input = 2.354 (image pixel value) then Output = Resultant image (corresponding annotations according to the pixel regions are given).

Here an annotation represents as the color representations.This part is the main one in fuzzy inference system. The rule which is not defined linguistically can lead to wrong output.

Layer 3: In Layer 3 normalization of membership functions are performed. Membership functions given here are the targets in which each pixel can be stored.

Layer 4: In Layer 4 execution of the fuzzy rule is performed. Execution of rules with the membership function is taken place here.

Layer 5: In Layer 5 summing up the output of the layer 4th and produces the output for the ANFIS system is taken place. Extraction of the rules according to the test image selected is displayed in the output.

ANFIS training gives better results as it combines both the benefits of neural network and Fuzzy systems. ANFIS classifier gives better performance than other two algorithms namely BPN and Fuzzy classifier in terms of its accuracy and speed to produce the output. It takes less time to generate the result. ANFIS classifier gives accuracy rate of 90% on SAR ice classification. Experimental results of ANFIS classifier is given in the Table 7 to Table 13. ANFIS classifier has some limitations which are listed below.

a. It creates a new structure for every incoming input since it combines the neural network structure.
b. Assigning the targets is another problem in ANFIS classifier. These targets should be analyzed and identified according to the problem which we going to perform.
c. Time which is taken for training the pixels is higher.
d. Rules must be linguistically defined.
e. According to the back propagation error rate, weights should be change until no errors are found.

2.3.4. Proposed Adaptive Neuro Fuzzy Inference System (ANFIS) with Particle Swarm Optimization (PSO) Classifier

To overcome the limitations found in the above discussed classifier a new novel method have been developed for SAR ice classification. Proposed method involves optimization technique which is combined with the previous classifier called ANFIS. PSO is an optimization method which is used to search for global optimized solution but time must be uncertain. The work starts with randomly initialized population for every individuals.PSO operates on the social behavior of particle. The global best solution is derived by adjusting each individual position with respect to global best position of particle with the entire population. Each individual is adjusting by altering the velocity with its observation of the particle and by its experience in search space. According to fitness function the new local best position and global best position will be calculated [11]. There are some numerous advantages of ANFIS with PSO classifier during the experimental, these are:

a. Finding fitness function in terms of optimized value: In ANFIS both the neural network and fuzzy classifier are combined. Training process is worked as neural network technique in which targets has to be given to the network. In combination of ANFIS with PSO, PSO will find its fitness function by using its optimized value. Fitness function is further used as the target.
b. Structure: In ANFIS new inputs are added to its network system as its works as neural networks it creates a new structure for its every incoming input. PSO does not create any new structure for newly coming inputs. It is used to update the new one with the same structure itself.
c. Computational Time: PSO will optimize its input according to the computational time; it produces the fitness result according to its optimized value so that the large amount of time will be reduced when training ANFIS.

    d.   Error : PSO aims to find out optimized vales, when it gets combined with ANFIS there is no possibilities of errors in training a network.

    e.   Defining Rule: PSO's optimized fitness value will be given to the rule set for ANFIS system. There is no need to set the rules in a linguistic manner as seen in the fuzzy and ANFIS system. To define a rule to perform the classification, calling fitness out put itself enough. For example, if fitness (1) then output=result.

The summarization of steps performed with ANFIS with PSO is given below.

Step 1: Assigning Inputs: input is taken are the pixel values of an image. There is no need to assign the targets in proposed work.

Step 2: Parameter initialization: PSO training parameters are initialized such as number of iterations, number of particles, number of swarms etc. These are initialized with empty matrices. During the training process the values are gets loaded for each iteration.

Step 3: Fitness function: After the training process over the outputs are produced this is said to be fitness function. This can be assigned as a target to the ANFIS system.

Step 4: Rules: According to the fitness function rules are defined. The great benefit in defining a rule is, is no need to define a rule linguistically. Just calling of fitness function is enough.

For ex, rule can be defined as,

If Fitness = 5.235 (image pixel value)

Then output = result (within this annotations are specified)

Step 5: Output: According to this value the output of classified image will be displayed.

In this paper, PSO is added to the ANFIS at the stage of training. Pixel values are given at the fuzzification process in proposed classifier. PSO optimize all these pixel values of SAR ice image and rearrange the membership function called as fitness function (here the optimize value taken are the input pixels). It gives the fitness value result according to the time of execution. The results which are produced from the PSO are given to the fuzzy rule generation, in which efficient manner of defining a rules are taken place. This leads the system to easily generate the output in less time and accurate manner. Compare to all other three classifiers proposed method gives better accuracy rate of 93.33%. The experimental results for ANFIS with PSO classifier is given in the Table 7 to Table 13.

3. Results and Discussion

This paper proposes a new classification algorithm called ANFIS with PSO for SAR ice classification. It has also presented various filter methods, segmentation algorithms and classification algorithms. The comparative study between all these methods is performed and it predicts the best one in all the three stages. The results are presented in the following paragraphs.

3.1. Preprocesing Results

Preprocessing stage aims at removing speckle noise from SAR ice images. There are various filters are performed such as, Median Filter, Lee Filter, Enhanced Lee Filter, Frost filter, Enhanced frost filter, Kaun filter and Gamma filter are from standard image speckle filter category and Symlet, Coiflet, Haar and Daubechies from wavelet family. All these filters performances are evaluated using following metrics.

    Signal to noise ratio
    Mean Absolute Error
    Root Mean Square Error
    Universal Image Quality Index

    a.   Signal to Noise Ratio (SNR) is used to measure the level of original signal affected by the level of back ground noise.

    b.   Mean Absolute Error (MAE) is used to measure how close the prediction to the actual outcome.

    c.   Root Mean Square Error (RMSE) is used to measure the difference between the prediction and observed outcome.

    d.   Universal Image Quality Index (UIQI) is used to measure the quality of an image after removing the noise. Table 5 shows the subjective results for the comparison of speckle filters.

Table 5. Subjective Results or Speckle Filter Comparison

Filters	Early winter sea ice, Beaufort				Islo Cedros			
	SNR	MAE	RMSE	UIQI	SNR	MAE	MSE	UIQI
Median Filter	5.682	7.145	12.457	0.874	8.987	8.244	8.254	0.874
Lee Filter	7.254	4.857	9.547	0.457	7.214	7.254	8.247	0.745
Enhanced Lee Filter	6.254	6.547	8.254	0.874	7.124	8.214	10.247	0.457
Frost Filter	4.257	7.654	7.254	0.598	8.578	6.214	7.024	0.789
Enhanced Frost Filter	4.896	9.354	7.547	0.578	8.254	9.254	5.456	0.457
Gamma Filter	8.257	5.254	8.574	0.754	3.254	5.254	7.254	0.459
Kaun Filter	6.578	8.254	7.254	0.874	4.256	6.547	11.657	0.687
Symlet Method	8.547	7.254	5.257	0.547	4.257	6.201	8.244	0.784
Coiflet Method	4.578	6.578	6.254	0.654	5.325	7.896	9.217	0.745
Haar Method	8.257	5.254	5.254	0.814	9.254	5.214	7.247	0.852
Daubechies Method	9.945	4.521	4.251	0.995	10.247	3.457	4.478	0.986

Filters	Grays harbor and willapa Bay				Mouth of Columbia river			
	SNR	MAE	RMSE	UIQI	SNR	MAE	RMSE	UIQI
Median Filter	7.254	7.245	12.248	0.204	5.245	5.328	6.215	0.457
Lee Filter	8.254	8.576	9.247	0.235	4.247	7.244	7.215	0.587
Enhanced Lee Filter	6.254	7.214	9.244	0.257	6.245	7.248	6.425	0.665
Frost Filter	6.257	6.247	5.241	0.368	5.247	8.245	6.430	0.228
Enhanced Frost Filter	5.254	8.247	8.242	0.345	8.254	9.247	10.412	0.784
Gamma Filter	6.254	8.248	7.245	0.456	7.289	9.244	8.825	0.578
Kaun Filter	5.241	9.248	10.247	0.745	9.235	9.254	7.252	0.874
Symlet Method	9.254	6.254	5.245	0.457	11.287	11.245	12.254	0.247
Coiflet Method	9.954	6.147	5.426	0.425	9.974	11.247	6.212	0.578
Haar Method	10.214	5.247	4.255	0.245	11.987	12.278	5.214	0.784
Daubechies Method	11.247	4.578	3.112	0.894	12.624	5.124	4.024	0.987

Filters	Strait of Georgia				Ward Hunt			
	SNR	MAE	RMSE	UIQI	SNR	MAE	RMSE	UIQI
Median Filter	7.247	4.275	9.254	0.579	5.272	8.287	9.382	0.124
Lee Filter	8.254	4.457	7.157	0.654	8.242	7.285	7.285	0.698
Enhanced Lee Filter	7.546	8.424	6.247	0.654	4.252	8.285	11.284	0.845
Frost Filter	6.242	4.248	7.257	0.789	7.472	9.385	7.285	0.174
Enhanced Frost Filter	8.245	5.452	8.254	0.657	7.552	7.284	12.524	0.213
Gamma Filter	5.245	8.427	9.475	0.694	7.282	8.274	5.965	0.125
Kaun Filter	4.575	7.427	7.851	0.378	6.274	9.281	9.382	0.368
Symlet Method	5.282	4.462	7.854	0.721	8.392	8.262	7.254	0.542
Coiflet Method	4.122	5.457	8.254	0.324	8.145	6.254	5.282	0.284
Haar Method	8.572	7.152	9.247	0.654	9.261	8.154	5.254	0.896
Daubechies Method	9.204	3.757	5.214	0.854	10.235	5.251	4.282	0.964

SNR- Signal to Noise Ratio, MAE- Mean Absolute Error, RMSE- Root Mean Square Error, UIQI- Universal Image Quality Index.

Daubechies method uses overlapping windows, thus the results reflects changes between its pixel intensities. Daubechies of D4 transform has four scaling coefficients. The sum of all these scaling coefficients is also one, so the calculation is depended over four adjacent pixels. Subjective results have shown that Daubechies from wavelet family has given a low error rate in MAE and RMSE parameters and higher rate in SNR, Universal Image Quality Index (UIQI) parameters.

3.2. Segmentation Results

Segmentation stage aims at separating the image into various regions. This paper aims at classifying the SAR ice images to know the regions around it. Thus segmentation includes dividing the images into four regions using K-means and Watershed algorithms. The area error segmentation metrics are used to evaluate the performance of these two algorithms is as follows:

  a. True Positive Area Ratio
  b. False Positive Area Ratio
  c. Similarity Index (SI)

Table 6 Shows the subjective results for comparison of K-means and Watershed segmentation algorithms, True Positive area ratio should be higher and False Positive area ratio should be lower and Similarity Index ratio should be higher in best segmentation algorithm. K-means has proven best in all these evaluation metrics.

Table 6. Subjective Results for K-means and Watershed Comparison

No. of Images	K-means			Watershed		
	TP in mm	FP in mm	SI in mm	TP in mm	FP in mm	SI in mm
1	17.2	8.9	16.5	9.1	16.2	8.1
2	15.2	8.8	14.2	8.9	15.1	7.8
3	17.5	7.7	16.5	8.9	17.3	6.7
4	16.2	7.6	15.8	8.9	16.1	6.6
5	16.9	8.2	16.5	8.9	16.7	7.2
6	16.5	6.3	14.2	8.9	15.2	5.3

TP- True Positive area ratio, FP- False Positive area ratio, SI- Similarity Index

3.3. Classification Results

Classification aims in identifying the regions in SAR ice images. The regions which are extracted can be identified as ice, water, vegetation and land. These areas can be represented in color options to know the different between the regions. There are four supervised classifiers such as BPN, Fuzzy classifier, ANFIS and ANFIS with PSO is considered in classification stage. All these classifiers performances are evaluated through following metrics.

  a. Accuracy
  b. Validity Index
  c. Sensitivity
  d. Time Complexity

Table 7 shows the subjective results for these classification algorithms.

Table 7. Subjective Results for these Classification Algorithms

Classifiers	Accuracy in %	Validity Index in %	Sensitivity in %	Time in Sec
BPN	77.14	0.0514	4.8913	6.11827
Fuzzy	86.6	0.0867	-	5.30179
ANFIS	90	0.0900	5.6604	3.54808
ANFIS +PSO	93.33	0.0933	5.8577	0.27243

Accuracy rate is used to define the overall performance of the classifiers; Validity Index is used to measure the performance of the network. Sensitivity is used to calculate the sensitivity on the parameters when it changes during training. It can be applied for BPN, ANFIS and ANFIS with PSO classifiers since all these classifiers uses network parameters to train a network, it is not suitable for fuzzy classifier since it does not uses parameters for training. Time complexity measure is used to predict the overall time has taken by the classifiers during the training.

Among all other classifiers ANFIS with PSO classifier gives higher rate in Accuracy, Validity Index, and Sensitivity and the lower rate in time complexity. Following Tables shows the classification results for all the classifiers included in SAR ice classification.

Table 8. Subjective Results for these Classification Algorithms

Classified Image of Early winter sea ice, Beaufort using BPN, Fuzzy, ANFIS and ANFIS with PSO

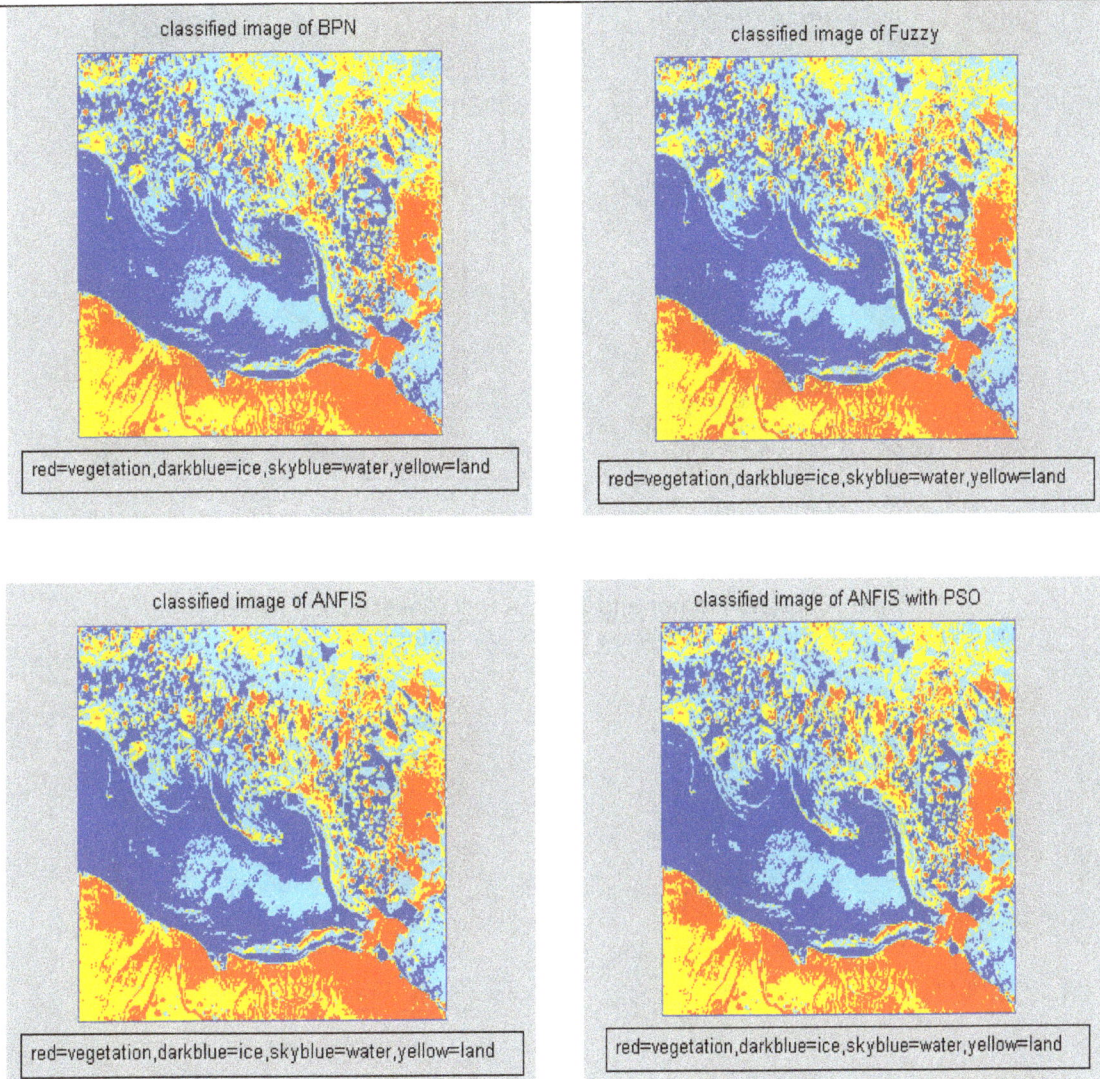

classified image of BPN

red=vegetation,darkblue=ice,skyblue=water,yellow=land

classified image of Fuzzy

red=vegetation,darkblue=ice,skyblue=water,yellow=land

classified image of ANFIS

red=vegetation,darkblue=ice,skyblue=water,yellow=land

classified image of ANFIS with PSO

red=vegetation,darkblue=ice,skyblue=water,yellow=land

Table 9. Islo Cedros Clasification Result

Classified Image of Islo cedros using BPN, Fuzzy, ANFIS and ANFIS with PSO

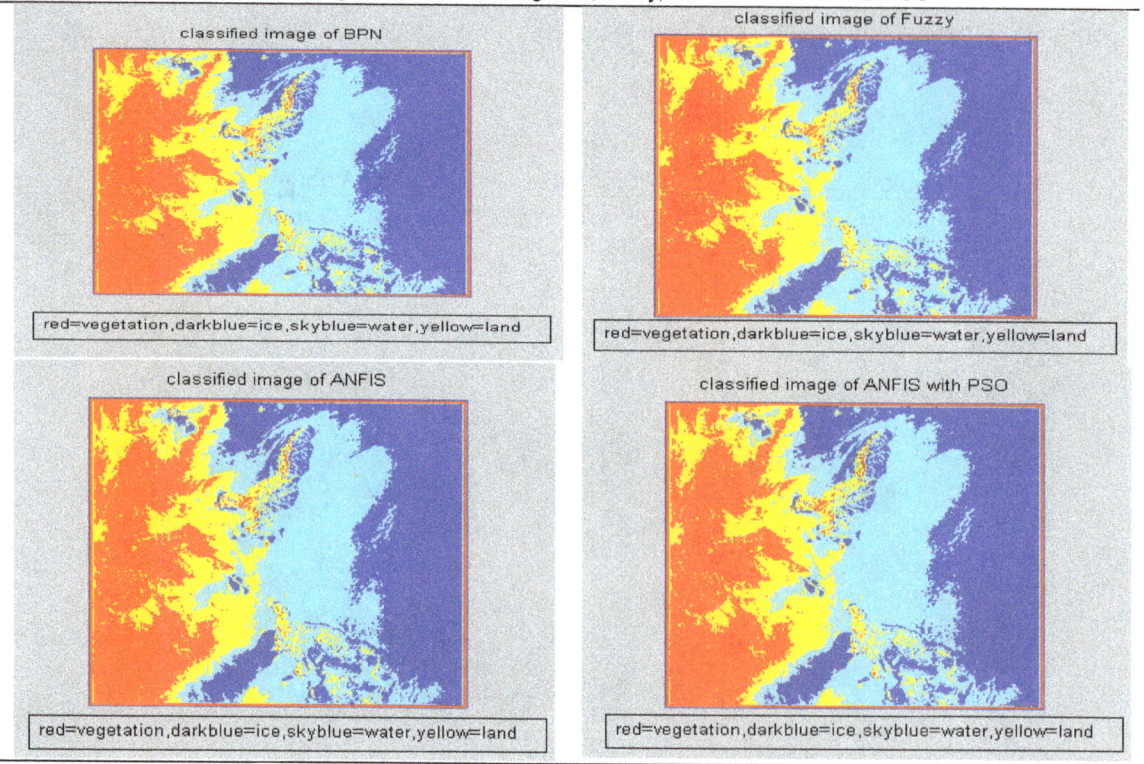

classified image of BPN

red=vegetation,darkblue=ice,skyblue=water,yellow=land

classified image of Fuzzy

red=vegetation,darkblue=ice,skyblue=water,yellow=land

classified image of ANFIS

red=vegetation,darkblue=ice,skyblue=water,yellow=land

classified image of ANFIS with PSO

red=vegetation,darkblue=ice,skyblue=water,yellow=land

Table 10. Grays Harbor and Willapa Bay Classification Result

Classified Image of Grays Harbor And Willapa Bay using BPN, Fuzzy, ANFIS and ANFIS with PSO

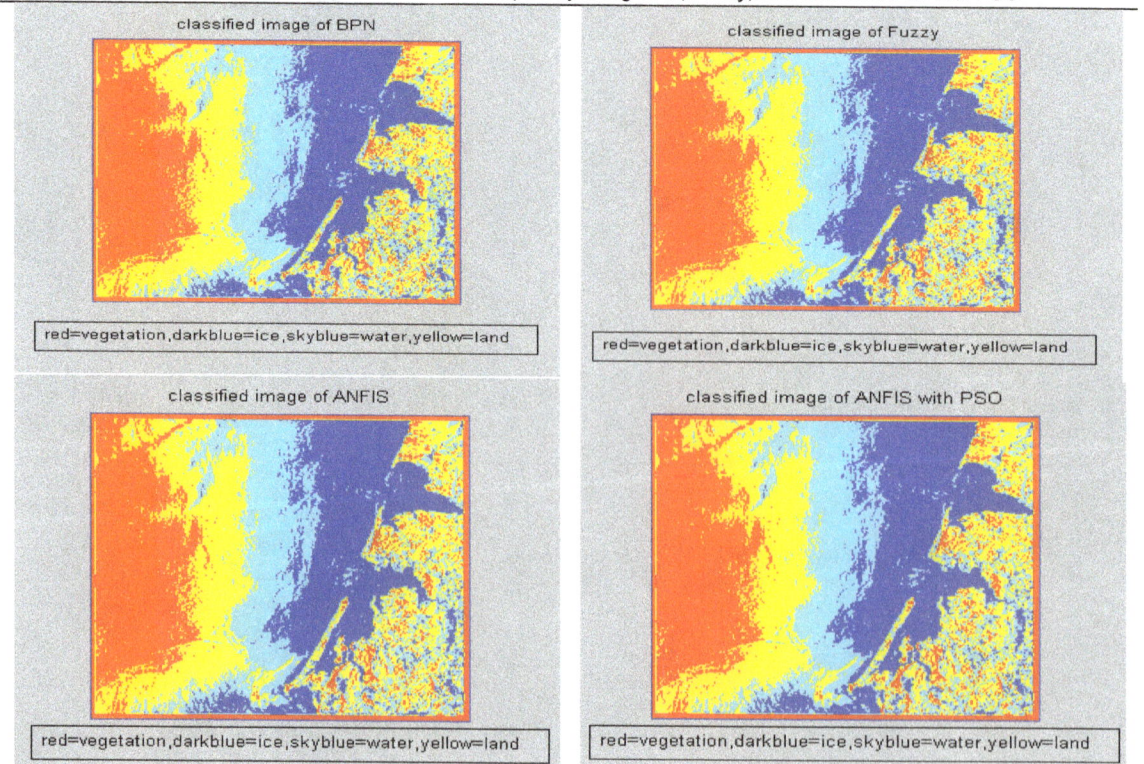

classified image of BPN

red=vegetation,darkblue=ice,skyblue=water,yellow=land

classified image of Fuzzy

red=vegetation,darkblue=ice,skyblue=water,yellow=land

classified image of ANFIS

red=vegetation,darkblue=ice,skyblue=water,yellow=land

classified image of ANFIS with PSO

red=vegetation,darkblue=ice,skyblue=water,yellow=land

Table 11. Mouth of Columbia River Classification Result

Classified Image of Mounth of Columbia River using BPN, Fuzzy, ANFIS and ANFIS with PSO

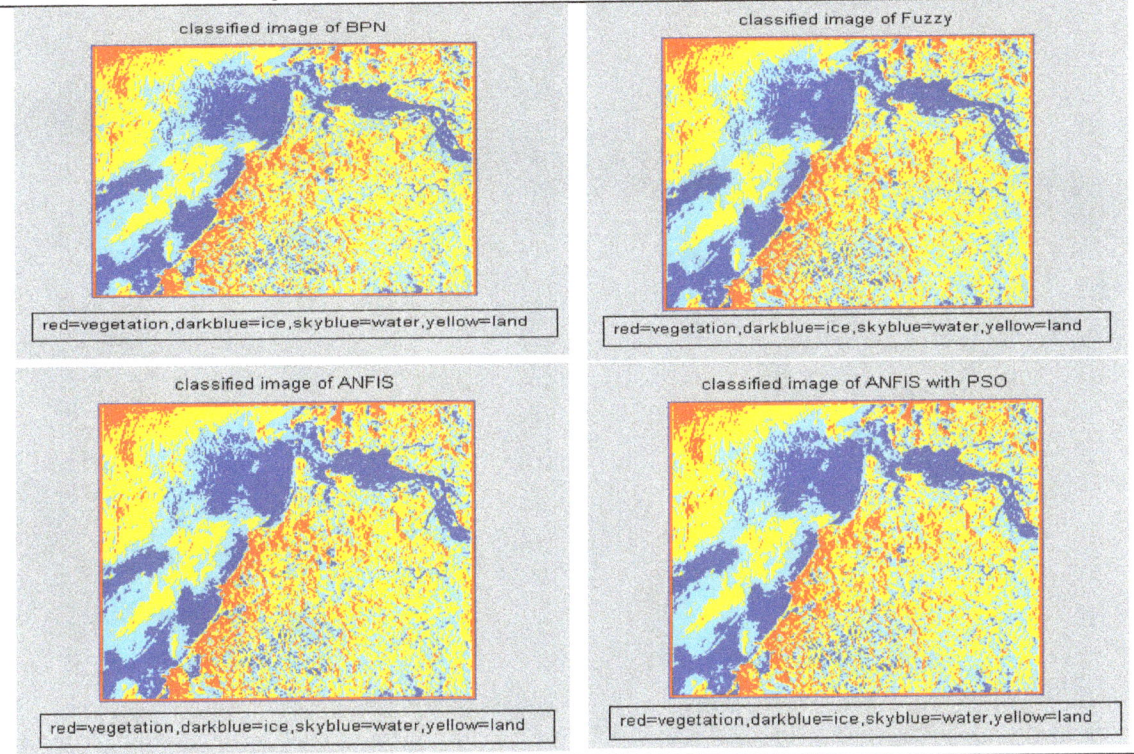

Table 12. Strait of Georgia Classification Result

Classified Image of Strait of Georgia using BPN, Fuzzy, ANFIS and ANFIS with PSO

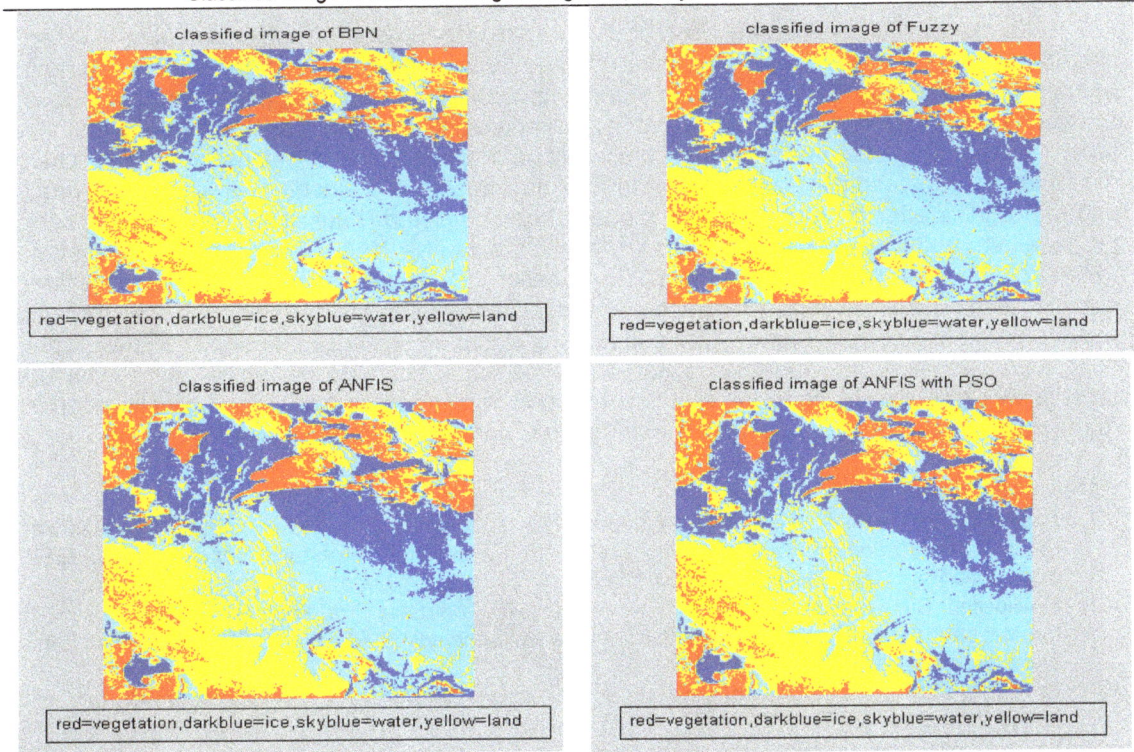

Table 13. Ward Hunt Classification Result

Classified Image of Ward Hunt using BPN, Fuzzy, ANFIS and ANFIS with PSO

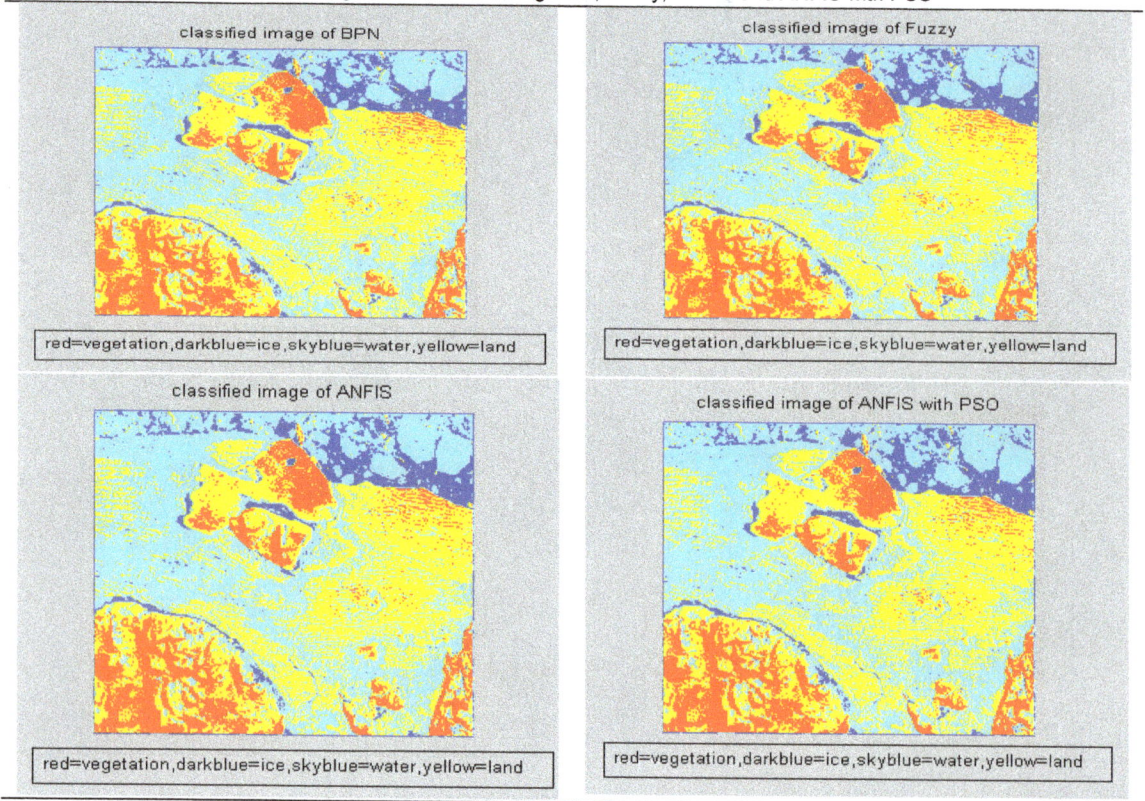

classified image of BPN

red=vegetation,darkblue=ice,skyblue=water,yellow=land

classified image of Fuzzy

red=vegetation,darkblue=ice,skyblue=water,yellow=land

classified image of ANFIS

red=vegetation,darkblue=ice,skyblue=water,yellow=land

classified image of ANFIS with PSO

red=vegetation,darkblue=ice,skyblue=water,yellow=land

4. Conclusion

In this paper, pixel based classification on SAR ice images using supervised learning classifiers are performed. SAR images are affected by speckle noise due to its backscattering nature of radar signals. This work is initialized with Denoising a speckle noise images using a Daubechies method in wavelet filter with the comparison of all the speckle filters. After the preprocessing stage segmentation is preceded. Comparison of K-means and Watershed algorithm are done and K-means is selected as a best segmented algorithm. Image was divided into the regions after the segmentation process has completed. Four classifiers namely, back propagation neural network (BPN), Fuzzy Classifier, Adaptive Neuro Fuzzy Inference System (ANFIS), and ANFIS with Particle Swarm Optimization (PSO) are performed. Above classifiers are the single classifier used to compare the accuracy of classification in SAR ice. Various regions called Vegetation, Land, water and ice are extracted by these classifiers. In this paper only four classifiers are performed for SAR ice classification, in future, fusion based classification can also be considered. It was focused on performing pixel based classification in future objective based classification can perform in fusion based classification.

References

[1]   Natalia Yu. Zakhvatkina, Vitaly Yu. Alexandrov, Ola M. Johannessen, Stein Sandven, and Ivan Ye. Frolov Classification of Sea Ice Types in ENVISAT Synthetic Aperture Radar Images. *IEEE Transactions on Geoscience And Remote Sensing*; 51.

[2]   Liu Huiying, Guo Huadong, Zhang Lu. *Sea Ice Classification using Dual Polarization SAR Data*. 35th International Symposium on Remote Sensing of Environment (ISRSE35) IOP Conf. Series: Earth and Environmental Science. 2014; 17.

[3]   S. Parrilli, M. Poderico, C.V. Angelino, G. Scarpa, and L. Verdoliva. A non local approach for SAR image Denoising. *IEEE*. 2010.

[4]   Klogo Griffith S., Gasonoo Akpeko and Ampomah K. E. Isaac. On The Performance of Filters for Reduction of Speckle Noise In SAR images off the Coast Of The Gulf of Guinea. *International Journal of Information Technology, Modeling and Computing (IJITMC)*. 2013; 1(4).

[5] Miao Ma, Jianhui Liang, Min Guo, Yi Fan, Yilong Yin. SAR Image Segmentation based on Artificial Bee Colony Algorithm. *Elsevier: Applied Soft computing*. 2011; 11(8): 5205–5214.

[6] Ashraf A. Aly, Safaai Bin Deris and Nazar Zaki. Research Review for Digital Image Segmentation Techniques. *International Journal of Computer Science & Information Technology (IJCSIT)*. 2011; 3(5).

[7] Pushpalata Pujari, Jyoti Bala Gupta. Improving Classification Accuracy by Using Feature Selection and Ensemble Model. *International Journal of Soft Computing and Engineering (IJSCE)*. 2012.

[8] Vidushi Sharma, Sachin Rai and Anurag Dev. A Comprehensive Study of Artificial Neural Networks. *IJARCSSE*. 2012.

[9] Vini Malik, Aakanksha Gautam, Aditi Sahai, Ambika Jha, Ankita Ramvir Singh. Satellite Image Classification Using Fuzzy Logic. *International Journal of Recent Technology and Engineering (IJRTE)*. 2013.

[10] V.Seydi Ghomsheh, M. Aliyari Shoorehdeli, M. Teshnehlab. *Training ANFIS Structure with Modified PSO Algorithm*. Mediterranean Conference on Control & Automation. 2007.

[11] Sarita Mahapatra, Alok Kumar Jagadev, Bighnaraj Naik. Performance Valuation of PSO Based Classifier for Classification of Multidimensional Data with Variation of PSO Parameters and Knowledge Discovery Database. *International Journal of Advanced Science and Technology*. 2011.

[12] Shuhrat Ochilov and David A. Clausi. *Automated Classification of Operational SAR Sea Ice Images*. Canadian Conference Computer and Robot Vision. IEEE. 2010: 978-0-7695-4040-5.

[13] Huawu Deng and David A. Clausi. Unsupervised Segmentation of Synthetic Aperture Radar Sea Ice Imagery using a Novel Markov Random Field model. *IEEE Transaction*. 2005; 43(3).

[14] S. Parrilli, M. Poderico, C.V. Angelino, G. Scarpa, L. Verdoliva. A Non Local Approach for SAR Image Denoising. *IEEE IGARSS: IEEE Geoscience and Remote Sensing Society*. 2010.

[15] K. Thangavel, R. Manavalan, Laurence Aroquiaraj. Removal of Speckle Noise from Ultrasound Medical Image based on Special Filters: Comparative Study. *ICGST GVIP: Journal Emphasizes on Graphics, Vision and Image Processing*. 2009; 9(3); 25-32.

[16] Klogo Griffith S, Gasonoo Akpeko, Ampomah K. E. Isaac. On the Performance of Filters For Reduction Of Speckle Noise in SAR Images Off the Coast of the Gulf Of Guinea. *IJITMC: International Journal of Information Technology, Modeling and Computing*. 2013; 1(4).

[17] Bala Prakash, Venu babu, Venu Gopal. Image Independent Filter for Removal of Speckle Noise. *IJCSI International Journal of Computer Science Issues*. 2011; 8(5).

[18] Jyoti Jaybhay, Rajveer Shastri. A Study of Speckle Noise Reduction Filters. *Signal & Image Processing: An International Journal (SIPIJ)*. 2015; 6(3).

[19] Asli Ozdarici, Zuhal Akyurek. *A Comparison of SAR Filtering Techniques on Agricultural Area Identification*. ASPRS: American Society of Photogrammetric and Remote Sensing Annual Conference. San Diego. 2010.

[20] Jaspreet kaur, Rajneet Kaur. Biomedical Images Denoising Using Symlet Wavelet with Wiener Filter. *International Journal of Engineering Research and Applications (IJERA)*. 2013; 3.

[21] Meenakshi Cahudary, Anupama Dhamija. A Brief Study of Various Wavelet Families and Compression Techniques. *Journal of Global Research in Computer Science*. 2013; 4(4).

[22] Sangeeta Arora, Yadwinder S. Brar, Sheo Kumar. Haar Wavelet Transform for Solution of Image Retrieval. *International Journal of Advanced Computer and Mathematical Sciences*. 2014; 5(2).

[23] Mohamed I. Mahmoud, Moawad I. M. Dessouky, Salah Deyab and Fatma H. Elfouly. Comparison Between Haar and Daubechies Wavelet Transforms on FPGA Technology. *International Journal of Electrical, Computer, Energetic, Electronic and Communication Engineering*. 2007; 1(2).

[24] Ashraf A. Aly, Safaai Bin Deris, Nazar Zaki, Research Review for Digital Image Segmentation Techniques. *International Journal of Computer Science & Information Technology (IJCSIT)*. 2011; 3(5).

[25] K.M. Sharavana Raju, Dr. V. Karthikeyani. Improved Satellite Image Preprocessing and Segmentation using Wavelets and Enhanced Watershed Algorithm. *International Journal of Scientific & Engineering Research*. 2012; 3(10).

[26] Sim Heng Ong, Kelvin W.C Foong, Poh-Sun Goh, Wieslaw Nowinski. *Medical Image Segmentation Using K-Means Clustering and Improved Watershed Algorithm*. IEEE Research Gate Conference paper. 2001.

[27] Johann Schumann, Yan Liu. Performance Estimation of a Neural Network-based Controller.

[28] Stein Sandven, Ola M. Johannessen. *Ers-L Ice Monitoring of The Northern Sea Route*. Proceedings of the First ERS-1 Pilot Project Workshop. Toledo, Spain. 1994.

20

Comparison Analysis of Model Predictive Controller with Classical PID Controller for pH Control Process

V.Balaji*[1], Dr. L.Rajaji[2], Shanthini K.[3]
[1]Department of Electrical Engineering, Singhania University, Pacheri Bari, Rajasthan, India
[2,3] ARM College of Engineering and Technology, MaraiMalai Nagar, Chennai, India
e-mail: balajieee79@gmail.com

Abstract

pH control plays a important role in any chemical plant and process industries. For the past four decades the classical PID controller has been occupied by the industries. Due to the faster computing technology in the industry demands a tighter advanced control strategy. To fulfill the needs and requirements Model Predictive Control (MPC) is the best among all the advanced control algorithms available in the present scenario. The study and analysis has been done for First Order plus Delay Time (FOPDT) model controlled by Proportional Integral Derivative (PID) and MPC using the Matlab software. This paper explores the capability of the MPC strategy, analyze and compare the control effects with conventional control strategy in pH control. A comparison results between the PID and MPC is plotted using the software. The results clearly show that MPC provide better performance than the classical controller.

Keywords*: pH control, PID, MPC, FOPDT, matlab*

1. Introduction

The Process control theory is evolving and new types of controller methods are being introduced.PID controllers arecommonly used due to its simplicity and effectiveness. In most of the industries PID controllers are used. Still there is no generally accepted design method for this controller. In 1970's MPC controller has been introduced as a way of controlling a wide range of processes. Now a day"s control systems engineers in the industry are adopting computer aided control systems design for modeling, system identification and estimation. These made a path to study MATLAB software .By adopting simulations the students may easily visualize the effect of adjustingdifferent parameters of a system and the overall performance of the system can be viewed. In this paper itis demonstrated how to create a model predictive controlfor a first order system with time delay in a MATLAB Simulink and also explains the difference betweenMPC and conventional controller.pH control plays a vital role in the process industry.The traditional method is to use classical PID method and the advanced control strategy includes ModelPredictive Controller. In this paper the tuning has beendone using Z-N Method and results have been compared between, PID and Model Predictive method.

2. Model Predictive Control

Model predictive control (MPC) has become a standard technology in the high level control of chemical processes. MPC or receding horizon control is a form of control in which the control action is obtained by solving on-line, at each sampling instant, a finite open-loop optimal control problem, using the current state of the plant as the initial state; the optimization yields an optimal control sequence in which the first control move is applied to the plant.

Here the controller tries to minimize the error between predicted and the actual value over a control horizon and the first control action is being implemented. Model predictive controllers rely on dynamic models of the process, most often linear empirical models obtained by system identification. MPC is also referred to as receding horizon control or moving horizon control (Qin and Badgwell, 2003).[3]

Figure 1 shows the behavior of an MPC system can be quite complicated, because the control action is determined as the result of the online Optimization problem. The problem is constructed on the basis of a process model and process measurements. Process measurements provide the feedback (and, optionally, feed-forward) element in the MPC structure. Figure 1 shows the structure of a typical MPC system feed-forward) element in the MPC structure.

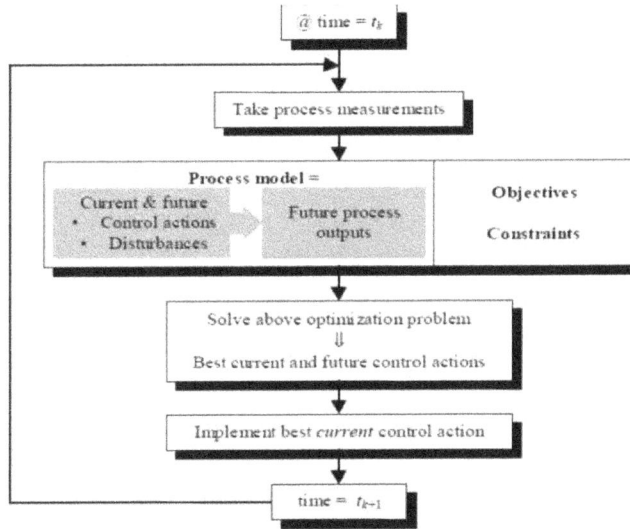

Figure 1. Model Predictive control Scheme

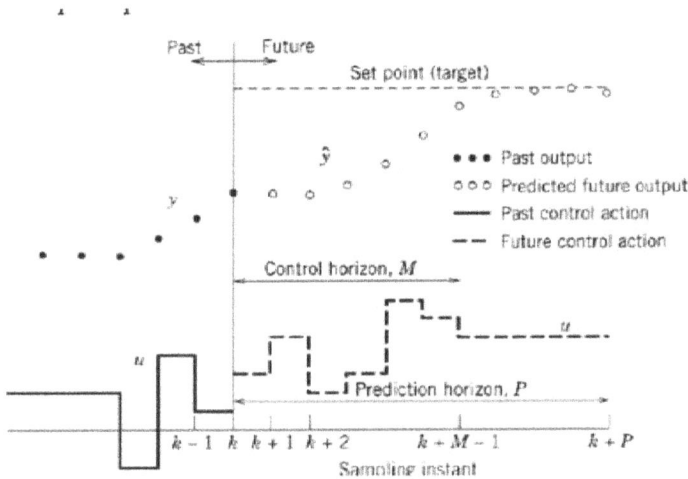

Figure 2. Basic concept of MPC [1]

2.1. The Receding Horizon

The control calculations are based on future predictions as well as current measurements. Future values of output variables are predicted using a dynamic model of the process and current measurements. Fig. 2 shows the concept of prediction horizon and control horizon.

2.2. Prediction and Control Horizons

Prediction horizon has a length equal to the number of samples in future for which the MPC controller predicts the plant output [1]. Prediction (P) is typically as far ahead as two to three times the dominant time constant of the system. Suppose the process is sampled at say one twentieth of that time constant: the output prediction horizon could then be up to some 60 steps ahead [2]. The length of control horizon is equal to the number of samples within the prediction horizon where the MPC controller can affect the control action [1].

2.3. Receding Horizon Approach

(i)At the kth sampling instant, the values of the manipulated variables, u, at the next M sampling instants, {u(k), u(k+1), ..., u(k+M -1)} are calculated. (ii)This set of M ―control moves‖ is calculated so as to minimize the predicted deviations from the reference trajectory over the next P sampling instants while satisfying the constraints. (iii) Typically, an LP or QP problem is solved at each sampling instant. Then the first ―control move‖, u(k), is implemented. (iv)At the next sampling instant, k+1, the M-step control policy is re-calculated for the next M sampling instants, k+1 to k+M, and implement the first control move, u(k+1).(v) Then Steps 1 and 2 are repeated for subsequent sampling instants.

3. Experimental Setup

The pH process is adjusted by controlling the flow rate of ammonia. This action adjusts the flow rate of the Ammonia, thus the input to the controller is the pH reading of the mixing vessel which is compared against the required set point. At the same time the output voltage obtained from the controller is used to adjust the solenoid valve or motorized valve to control the Ammonia flow rate. This output tends to maintain the mixing vessel pH to a desired value. The Figure 3 shows the pH controlling process.

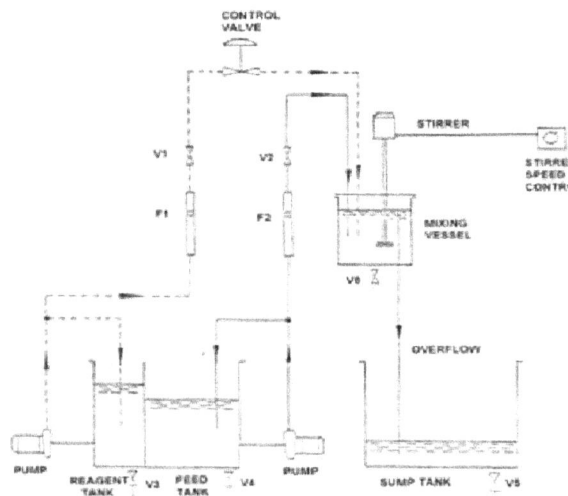

Figure 3. Process diagram of pH control

3.1. Approximating pH process to FOPDT

For the step input of (0 to 30% opening of Ammonia flow rate valve), we note the following characteristics of its step response: to approximate into First Order Plus Delay Time (FOPTD) model,

(i)   The response attains 63.2% of its final response at time, t = τ+θ. (ii) The line drawn tangent to the response at maximum slope (t = θ) intersects the y/KM=1 line at (t = τ+ θ). (iii) The step response is essentially complete at t=5τ. In other words, the settling time is ts=5t. The graphical the analysis to determine the FOPDT model is shown in the fig 4. Therefore the FOPDT model transfer function becomes

$$TF = \frac{4.67e^{-0.96s}}{0.4s + 1}$$

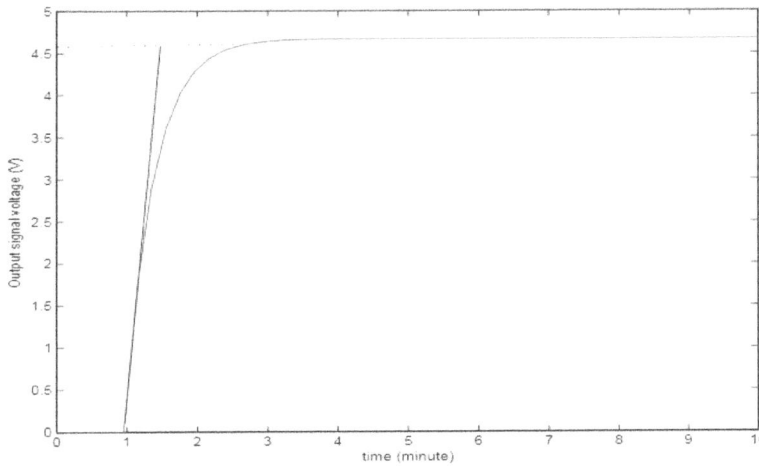

Figure 4. Graphical Analysis to Obtain the Model

4. Methodology

The Design of conventional PID controller and advanced controller are done using the Matlab tools. Figure 5 and 6 shows the diagram of PID and MPC tuning in the Matlab environment.The setpoints needed to be adjusted are 1.8,1.9,2.8 and 4.8.The PID controller gives good setpoint tracking when kp= 0.10, τI=0.17 and τd=0.033. The MPC is tuned prediction horizon of 3, control horizon of 1 and control interval of 1.

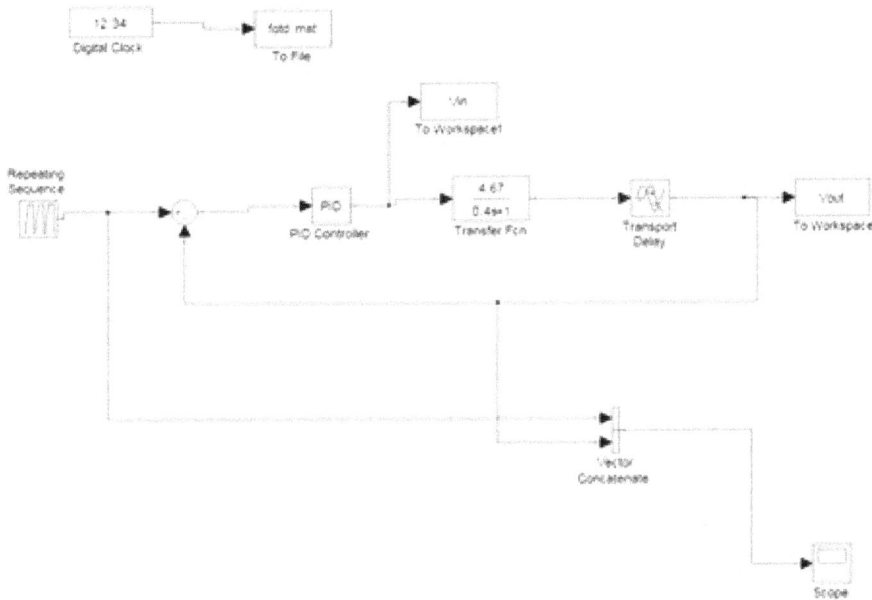

Figure 5. Simulink Diagram for PID Design

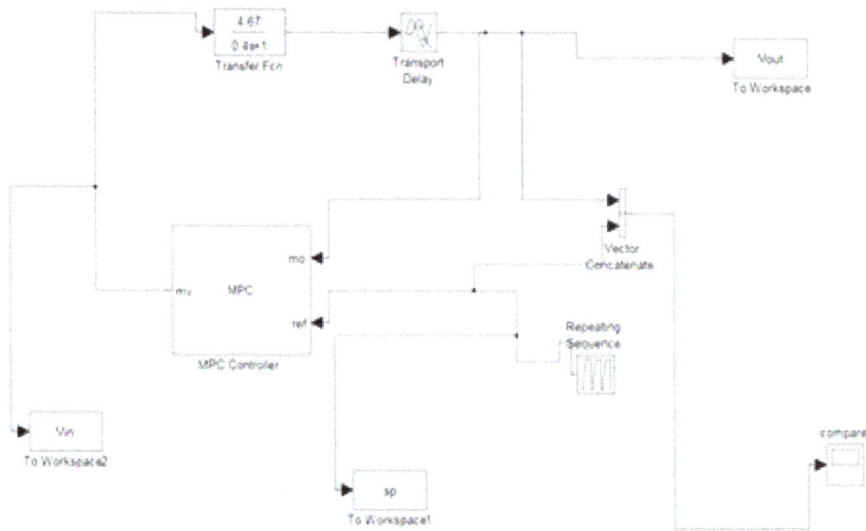

Figure 6. Simulink Diagram for MPC Design

5. Results and Analysis

The graph between time and the output signal has been obtained for PID, and MPC controller as shown in Figure 7. The comparison between these controllers has been done and the best controller has been obtained.

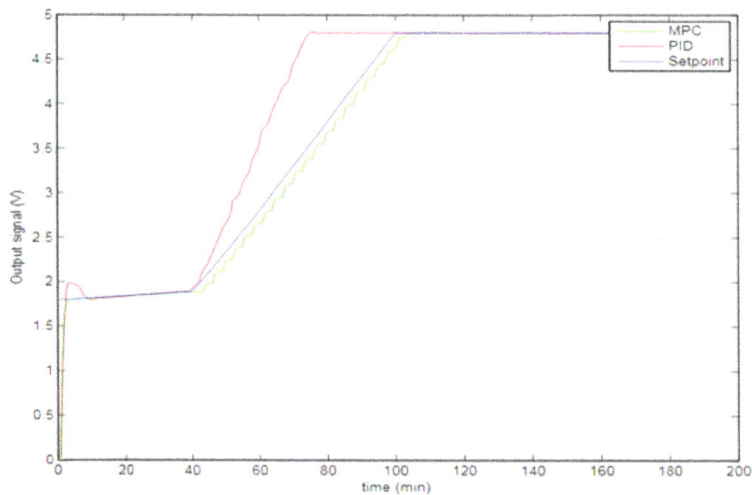

Figure 7. Output Response of PID and MPC Controller

We can observe from Figure 7 that how fast the MPC can reach the set-point. In the response of the PID we can easily the fulucations from the begining itself and it is time consuming to reach the set point. Figure 8 shows the graph of input adjustment for both the controllers.

The output response of the MPC is faster than the response of the PID controller.

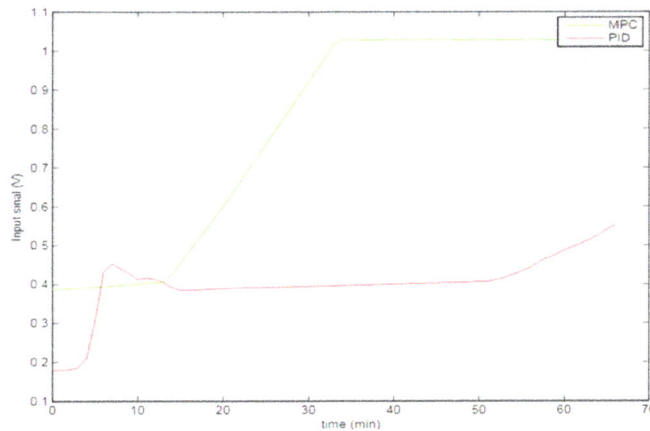

Figure 8. Input Adjustments by MPC and PID

6. Conclusion

The obtained transfer function is processed using classical and advanced controllers such as PID, and MPC. The values which are obtained from the tuning methods are simulated using MATLAB. It is seen from the response curve that MPC controller provides a better response with minimum time when compared with PID and IMC. So it is concluded that MPC controller is efficient for a pH process.

References

[1] Hans-Petter, Halvorsen. Model Predictive Control in labVIEW. *Dept. of Electrical Engineering Information Technology and Cybernetics*, Telemark University College. 2011.
[2] Jonathan L. Process Automation Handbook - A guide to Theory and Practice.
[3] Bemporad, A. Model Predictive Control: Basic concepts, Controllo di Processo e Sistemi di Produzione. 2009.
[4] Camacho and Bordons. Model Predictive Control. Springer. 2004.
[5] Rossiter J. Model based Predictive Control - A practical approach. CRC Press. 2003.
[6] Mathworks.com. Model Predictive Control Toolbox. 2012.
[7] J.Prakash, K.Srinivasan. Design of Non Linear PID Controller and Non Linear Model Predictive Controller for a Continuous stirred tank Reactor. ISA Transactions. 2009; 48: 273-282.
[8] S.Abirami, H.Kala, P.B.Nevetha, B.Pradeepa, R.Kiruthiga, P.Sujithra. Performance Comparison Of Different Controllers For Flow Process. *International Journal of Computer Applications*. 2014; 90(19): 17-21.
[9] E.F. Camacho, C. Bordons. Model Predictive Control in the Process. Springer-verlay. 1995.
[10] Ljung. System Identification: Theory for the User. Printice-Hall: Englwood cliffs. 1987

21

A Pattern Classification based Approach for Blur Classification

Shamik Tiwari
CET, Mody University of Science & Technology, Lakshmangarh
e-mail: shamiktiwari@hotmail.com

Abstract

Blur type identification is one of the most crucial steps of image restoration. In case of blind restoration of such images, it is generally assumed that the blur type is known prior to restoration of such images. However, it is not practical in real applications. So, blur type identification is extremely desirable before application of blind restoration technique to restore a blurred image. An approach to categorize blur in three classes namely motion, defocus, and combined blur is presented in this paper. Curvelet transform based energy features are utilized as features of blur patterns and a neural network is designed for classification. The simulation results show preciseness of proposed approach.

Keywords: *Blur, Motion, Defocus, Curvelet Transform, Neural Network.*

1. Introduction

Barcoding is one of the automatic identification and data collection (AIDC) technology which reduces human involvement in data entry and collection and in that way also dropping error and time in the process. These days, a wide range of general purpose hand-held devices, such as mobile phones, come with an optical imaging system. Enabling these general purpose hand-held devices with the capability to recognize barcodes is a cost-effective alternative to conventional barcode scanners [1, 2]. The availability of imaging phones provides people a mobile platform for decoding barcode rather than the use of the conventional scanner which has lack of mobility.

Unfortunately, sometimes the deprived quality of the images taken by digital cameras makes it difficult to correctly decode barcodes. Barcode images are usually prone to number of degradations. Image blurring is often a major factor influencing the performance of a barcode recognition system. Image blurring is usually inevitable in a camera-based imaging system, especially in the case of the camera that does not have auto focus or macro mode [3,4]. Although some high-end camera phones which integrate a high resolution and auto focus/macro mode are available, the low-end camera phone user segment is still huge.

The objective of image restoration approach is to recover a true image from a degraded version. This problem can be stated as blind or non-blind depending upon whether blur parameters are known prior to the restoration process. Blind restoration deals with parameter identification before deconvolution. Though there exists multiple blind restorations techniques but blur type recognition is extremely desirable before application of any blur parameters estimation approach [5].

While most previous work focuses on image deblurring, not as much research has been done on blur classification, which is more practical because the type of blurs is usually unknown in images. Based on the descriptor of blurs, there are a few blur classification methods. Multilayer neural network is a neural network, which is based on the multi valued neurons (MLMVN) with traditional feed forward architecture. The MLMVN is used by Aizenberg et al. [6] to identify types of the blur, whose precise identification is essential for the image deblurring. Chong et al. [7] have proposed a scheme that simultaneously detects and identifies blur. This method is based on the analysis of extrema values in an image. The extrema of histograms are first identified then analyzed in order to extract feature values in this method. Liu et al. [8] have offered a framework for partial blur detection and classification i.e. whether some portion of the image is blurred as well as what type of blur arises. They considered maximum saturation of color, gradient histogram span and spectrum details as blur features. Aizenberg et al. [9] presented a work which identifies blur type, estimates blur parameters and perform image restoration using neural network. They considered four kinds of blur namely rectangular, motion, defocus, and Gaussian blurs as a pattern classification problem. Yan et al. [10] made an attempt to find a general feature extractor for common blur kernels with various parameters, which is closer to realistic application scenarios and applied deep belief neural networks for discriminative learning. In this method Fourier spectrum of blurred images is passed to the neural network as input. However, utilizing spectrum as an input is not a good idea due to large size of feature set. In the paper [11], Tiwari et al. use statistical features of blur patterns in frequency domain and identified blur type with the use of feed forward neural network. Tiwari et al. [12] used wavelet features for blur type recognition utilizing a neural network.

Curvelet features have been extensively used in object and texture classification. However, we have not found any application of curvelet features for blur classification. In this paper we extract curvelet features of blur patterns in frequency domain and use feed forward neural network for blur classification. This paper is organized into seven sections including the present section. In section two to four, we discuss the theory of image degradation model, curvelet transform and feed forward neural network in that order. Section five describes the methodology of blur classification scheme. Section six discusses experimental results and in the final section seven, conclusion is presented.

2. Image Degradation Model

The image degradation process of an image can be modelled as the following convolution process [12].

$$g(x, y) = f(x, y) \otimes h(x, y) + \eta(x, y) \qquad (1)$$

where $g(x, y)$ is the degraded image, $f(x, y)$ is the uncorrupted original image, $h(x, y)$ is the point spread function that caused the degradation and $\eta(x, y)$ is the additive noise in spatial domain. In view of the fact that, convolution operations used in spatial domain is equivalent to the multiplication in frequency domain, so image degradation model is

$$G(u, v) = F(u, v)H(u, v) + N(u, v) \qquad (2)$$

Motion blur occurs in an image due to relative motion between image capturing device and the object. Let the image to be acquired has a relative motion to the capturing device by a regular velocity (vrelative) and makes an angle of α radians with the horizontal axis for the duration of the exposure interval [0, texposure], the distortion is one dimensional. Expressing motion length as $L = v_{relative} \times t_{exposure}$, the point spread function (PSF) for uniform motion blur can be modeled as [12]:

$$h(x, y) = \begin{cases} \frac{1}{L} & \text{if } \sqrt{x^2 + y^2} \leq \frac{L}{2} \text{ and } \frac{x}{y} = -\tan \alpha \\ 0 & \text{otherwise} \end{cases} \qquad (3)$$

The defocus blur also known as out of focus blur is appears due to a system of circular aperture. It can be modeled as a uniform disk as [13]:

$$h(x, y) = \begin{cases} \frac{1}{\pi R^2} & \text{if } \sqrt{x^2 + y^2} \leq R \\ 0 & \text{otherwise} \end{cases} \qquad (4)$$

where R defines the radius of the disk.

Sometimes image contains co-existence of both blurs. In that case the blur model becomes [14]:

$$h(x, y) = a(x, y) \otimes b(x, y) \qquad (5)$$

where $a(x, y)$, $b(x, y)$ are point spread functions for motion and defocus blur respectively and $\otimes$ is the convolution operator. The degradation process model equations (1) and (2) can be expressed as (6) and (7) in spatial and frequency domain respectively:

$$g(x, y) = f(x, y) \otimes a(x, y) \otimes b(x, y) + \eta(x, y) \quad (6)$$

$$G(u, v) = F(u, v)A(u, v)B(u, v) + N(u, v) \qquad (7)$$

3. Discrete Curvelet Transform

Curvelet Transform by Candes and Donoho [15, 16] was designed to overcome the inherent limitations of traditional multiscale transforms like wavelet. Curvelet transform is a multi-scale and multi-directional transform with needle shaped basis functions. Basis functions of wavelet transform are isotropic and thus it requires large number of coefficients to represent the curve singularities. Curvelet transform basis functions are needle shaped and have high directional sensitivity and anisotropy. Curvelets obey parabolic scaling.The curvelet has a frequency support in a parabolic wedge area due to the anisotropic scaling law as width = length2.

The fast discrete curvelet transform was introduced by Candes et al. [17] in two forms, the wrapping version and the unequally spaced FFT (USFFT) version. Since the wrapping version is faster and invertible up to numerical precision, while the USFFT version is only approximately invertible, we use only the wrapping version throughout this paper. The two dimensional curvelet transform is given by

$$C^D(j, l, k) = \sum_{x=0}^{M-1} \sum_{y=0}^{N-1} f[x, y]\emptyset_{j,l,k}^D[x, y] \qquad (8)$$

where $f[x, y], 0 \leq x < M, 0 \leq y < N$ is the input 2-D image. $C^D(j, l, k)$ are the discrete curvelet coefficients, $\emptyset_{j,l,k}^D[x, y]$ is the curvelet basis function at scale j, orientation $l \in [0, 2\pi]$ and location $k[k_1, k_2]$.

4. Feed Forward Neural Network

A successful pattern identification method depends significantly on the particular selection of the classifier. An artificial neural network (ANN) is a system which can be seen as an information-processing paradigm. ANN has been designed as generalizations of mathematical models identical to human recognition system. They are collection of interconnected processing units called neurons that perform as a unit. It can be used to determine complex relationships among inputs and outputs by recognizing patterns in data. The feed forward neural network refers to the neural network which contains a set of source nodes which forms the input layer, one or more than one hidden layers, and single output layer. In case of feed forward neural network input signals propagate in one direction only; from input to output. There is no feedback path i.e. the output of one layer does not influence same layer. One of the best known and broadly acceptable learning algorithms in training of multilayer feed forward neural networks is Back-Propagation [18]. The back propagation is a type of supervised learning algorithm, which means that it receives sample of the inputs and associated outputs to train the network, and then the error (difference between real and expected results) is calculated [19]. The idea of the back propagation algorithm is to minimize this error, until the neural network learns the training data. This can be implemented by:

$$\Delta w_{ji}(n+1) = \eta(\partial_{pj}O_{pj}) + \alpha \Delta w_{ji}(n) \qquad (9)$$

where η is the learning rate, α is the momentum, ∂_{pj} is the error and n is the number of iteration.

5. Blur Classification Approach

Blur type can be identified in frequency domain notably. The motion blur, defocus blur are appeared differently in frequency domain, and the blur categorization can be easily done using these specific patterns. The frequency spectrum of motion blurred image has dominant dark parallel lines, which are orthogonal to the motion orientation with near zero values. On the other hand, In case of defocus blur one can see appearance of some circular zero crossing patterns and in case of coexistence of both blurs combined effect of both blurs become visible. Figure 1 shows the effect of different blurs on the Fourier spectrum of original image. We can consider these patterns in frequency domain as image itself for blur classification.

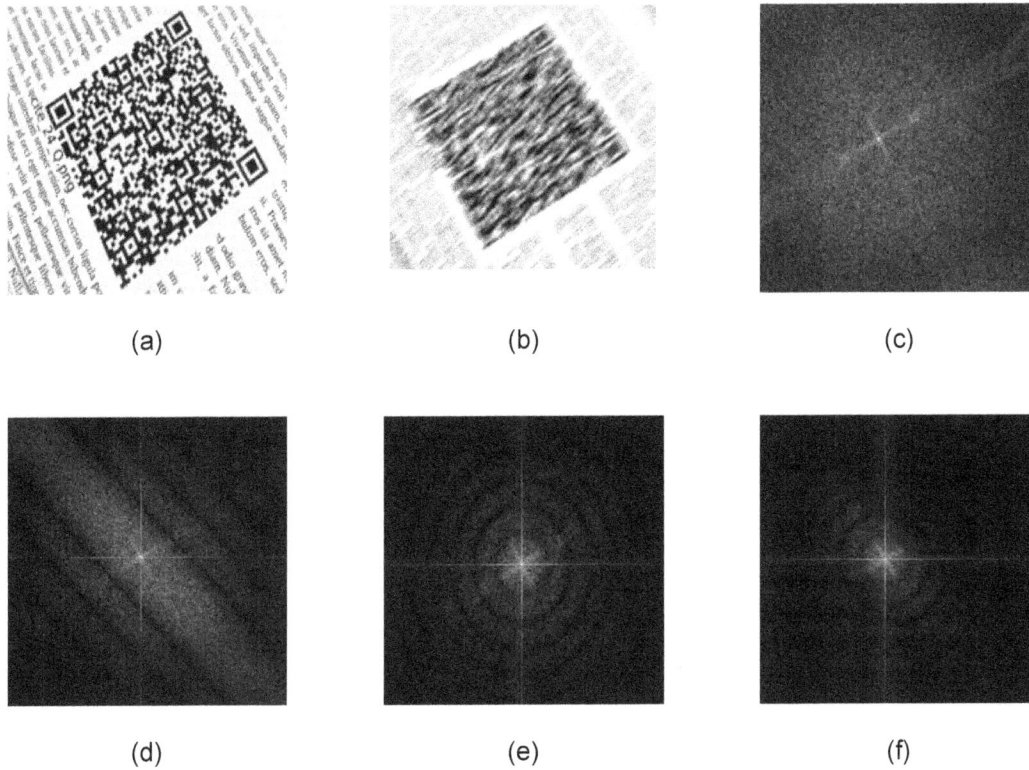

(a)   (b)   (c)

(d)   (e)   (f)

Figure 1(a) Original image containing QR code [18] (b) Motion blurred image
(c) Fourier spectrum of original image (d) Fourier spectrum of blurred image with motion
length 10 pixels and motion orientation 450 (e) Fourier spectrum of blurred image with defocus
blur of radius 5 and (f) Fourier spectrum of image with coexistence of both blurs

Main steps of the blur classification method are: image preprocessing, find logarithmic frequency spectrum, feature extraction of blur patterns, designing of neural network classifier system and result analysis.

In the initial step, we perform preprocessing on blurred images. Then we find the logarithmic frequency spectrum of blurred and non blurred image to get the blur patterns. Once the blur patterns acquired, the curvelet energy features have been calculated to prepare the training and testing database. Finally with this feature database training and testing are performed by using feed forward back propagation neural network.

5.1. Preprocessing
Blur classification requires a number of preprocessing steps. First, the color image obtained by the digital camera is changed into an 8-bit grayscale image by averaging the color channels.

The periodic transitions at the boundaries of an image often lead to high frequencies, which are transformed into visible vertical and horizontal lines in the power spectrum. Since, these lines may divert from or even superpose the stripes caused by the blur, they have to be detached by applying a windowing function before transformation. The Hanning window gives a fine trade-off between forming a smooth transition towards the image borders and keeping adequate image information in power spectrum. A 2-D Hann window of size N X M defined as the product of two 1-D Hann windows:

$$w[n, m] = \frac{1}{2}\left(1 + \cos[2\pi \frac{n}{N}]\right) . \frac{1}{2}\left(1 + \cos[2\pi \frac{m}{M}]\right) \qquad (10)$$

After that step, the windowed image is transformed to the frequency domain using Fourier transform. Then, the power spectrum is achieved. However, as the coefficients of the spectrum decrease rapidly from its centre to the borders, it is difficult to identify local differences significantly. This fast drop off is reduced by considering the logarithm of the power spectrum. In order to obtain a centered version of the spectrum, its quadrants have to be swapped diagonally. A centered portion of size 256 X 256 is cropped to perform feature extraction, in view of the fact that the significant features are around the centre of the spectrum,

5.2. Feature Extraction

Wrapping based discrete curvelet transform using Curvelab-2.1.2 is applied to a power spectrum of barcode image to obtain its coefficients. These coefficients are then used to form the features of blur patterns. After achieving the curvelet coefficients the mean and standard deviation of the coefficients related with each subband is calculated at the coarsest and the finest scale independently. The mean of a subband at scale j and orientation l is calculated as:

$$\mu_{j,l} = \frac{E(j, l)}{M \times N} \qquad (11)$$

where $M \times N$ is the size and $E(j, l)$ is the energy of curvelet transformed image respectively at scale j and orientation l. Energy is calculated by the sum of absolute values of curvelet coefficients.

$$E(j, l) = \sum_{x=0}^{M-1} \sum_{y=0}^{N-1} \left|C_{j,l}^D(x, y)\right| \qquad (12)$$

The standard deviation of a subband at can be shown as:

$$\sigma_{j,l} = \frac{\sqrt{\sum_{x=0}^{M-1} \sum_{y=0}^{N-1} \left(\left|C_{j,l}^D(x,y)\right| - \mu_{j,l}\right)^2}}{MXN} \qquad (13)$$

We have used 3 level discrete curvelet decomposition. Table 1 shows the subbands division at each level of transform.

Table 1. Curvelet transform with 3 levels

Scale	1	2	3
Total No. of subbands	1	16	32

Only first half of the total subbands at a scale are considered for feature calculation because in the frequency domain the curvelet at angle l produces the similar coefficients as the curvelet at angle $(l + \pi)$. Based on the subband division in table 1, (1+8+16) = 25 subbands of curvelet coefficients are selected for calculation of mean and standard deviation. All these features are arranged in a manner such that the standard deviations remain in the first half of the feature vector and the means are arranged into the second half of the feature vector. So, we obtain a feature vector consists of mean energies and standard deviations $f = (f_1, f_2 ..., f_{50})^T$ for each of the power spectrum. Where first twenty five values are standard deviation and others are mean energies. We use this feature vector f to classify blur patterns

6. Simulation Results

The performance of the proposed technique has been evaluated using camera based barcode images. To simulate the blur classification algorithm, we have considered two different barcode image databases. The first database WWU Muenster Barcode Database [20] consisting of 1-D barcode images and the second one is the Brno Institute of Technology QR code image database [21] captured by digital camera. This complete work is implemented using neural network and image processing toolbox of Matlab 6.5.

Most of pattern recognition problems are too complex to be solved completely by non automated algorithms, machine learning has always played a crucial role in this area. The conventional way of identification is to split the recognizer into a feature extractor, and a classifier. One of the main achievements of the neural network is to offer complex mapping in high-dimensional spaces without requiring complicated hand-crafted features. This allows designers to rely more on learning, and less on detailed engineering of feature extractors. These features of neural network enforced to use as a classification tool in this work.

The classification is achieved using a feed forward neural network containing single hidden layer fifty neurons. Back propagation is pertained as network training principle, where the training dataset is designed by the extracted features of the blur patterns. The whole training and testing features set is normalized into the range of [0, 1], whereas the output class is assigned to zero for the lowest probability and one for the highest. In this work the transfer functions of hidden layers are hyperbolic tangent sigmoid functions. The numbers of neurons used in input layer are equal to the extracted features from image dataset which are fifty. The neural classifier is trained with different number of hidden layers. The final architecture is selected with single hidden layer of fifty neurons which gives best performance.

To design the neural network, the data is divided into training, validation and test sets in a ratio of 50:20:30 respectively. Neural networks are known to be prone to overfitting, and hence not performing well on previously-unseen data. In order to reduce the risk of overfitting, a validation dataset was used to test each network periodically during training. When the validation set error began to rise, training was terminated. Only at that point the network was tested with the test data set in order to calculate the final accuracy. Three different partition sets were created using random partition. Partitions satisfied cross-validation, meaning that each target class (i.e. blur) was represented the same number of times in the training set, the same number of times in the validation set and the same number of times in the test set. For each partition, every pattern appeared in exactly one of the training, validation or test sets. Finally, we used overall classification results to evaluate the blur classification model.

When referring to the performance of a classification model, we are interested in the model's ability to correctly predict or separate the classes. When looking at the errors made by a classification model, the confusion matrix gives the full picture. In confusion matrix, the diagonal cells show the number of instances that were correctly classified for each structural class. The off-diagonal cells show the number of instances that were misclassified. To assess the blur classification model, we have shown confusion matrix and used the four statistical metrics as specified below.

Precision (Positive Predictive Value): Precision measures the fraction of samples that truly turns out to be positive in the group the classifier has categorized as a positive class. The higher the precision is, the lower the number of false positive errors committed by the classifier.

$$Precision = \frac{True\ Positive}{True\ Positive + False\ Positive}$$

Sensitivity (Recall): It measures the actual members of the class which are accurately recognized as such. It is also referred as true positive rate (TPR). It is defined as the fraction of positive samples categorized accurately by the classification model. High sensitivity value represents that low number of positive samples misclassified as the negative class.

$$Recall = \frac{True\ Positive}{True\ Positive + False\ Negative}$$

Specificity: Recall/sensitivity is related to specificity, which is a measure that is commonly used in multi class problems, where one is more interested in a particular class. Specificity corresponds to the true-negative rate.

$$\text{Specificity} = \frac{\text{True Negative}}{\text{True Negative} + \text{False Positive}}$$

Accuracy: Accuracy is the overall correctness of the model and is calculated as the sum of correct classifications divided by the total number of classifications.

Where the terms true positive, true negative, false positive and false negative are defined below.

(i) True Positive: In case of test pattern is positive and it is categorized as positive, it is considered as a true positive.

(ii) True Negative: In case of test pattern is negative and it is categorized as negative, it is considered as a true negative.

(iii) False Positive: In case of test pattern is negative and it is categorized as positive, it is considered as a false positive.

(iv) False Negative: In case of test pattern is positive and it is categorized as negative, it is considered as a false negative

Experiment 1: Performance analysis with blurred 1-D images

In the first experiment, we have considered first 200 images from the 1-D barcode database. Then, the three different classes of blur (motion, defocus and combined blur) were synthetically introduced with different parameters to make the database of 600 blurred images (i.e., 200 images with each class of blur) for each type of barcode images. Out of 600 blurred images, 300, 120 and 180 images are used as training, validation and testing datasets respectively to design the neural model. The training process is continued till the best validation performance achieved. The training stopped when the validation error increased for 6 iterations. The best validation performance is 0.0115, at epoch 39 as shown in Figure. 2

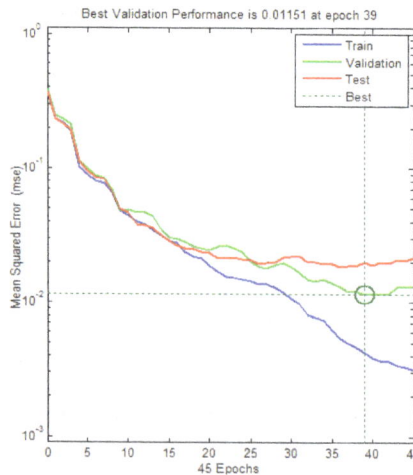

. Figure 2. Network Performance

Table 2. Confusion matrix for blur classification results of blurred 1-D barcode images

Predicted Class		Target Class		
		Motion Blur	Defocus Blur	Combined Blur
	Motion Blur	189	0	0
	Defocus Blur	11	200	0
	Combined Blur	0	0	200

Table 3. Blur Classification results for blurred 1-D barcode images

Blur Type	Precision	Recall	Specificity
Motion Blur	100	94.5	100
Defocus Blur	94.8	100	97.2
Combined Blur	100	100	100
Accuracy			98.2

The trained network model is used subsequently to categorize the blur type from degraded images. The overall blur classification results are presented as confusion matrix in Table 2. Using this confusion matrix statistical quantities precision, recall, specificity and accuracy are calculated. The high values of these metrics in Table 3 show effectiveness of our model. The overall classification accuracy is 98.2 %.

Experiment 2: Performance analysis with blurred and noisy 1-D images

To test the efficiency of the proposed scheme in presence of noise, another neural model is created using 200 1-D barcode images. All the images are degraded different kinds of blur similar to Experiment 1. Subsequently, Gaussian noise of 40dB Blurred Signal-to-Noise Ratio (BSNR) is added to the blurred images to create a database of blurred and noisy samples.

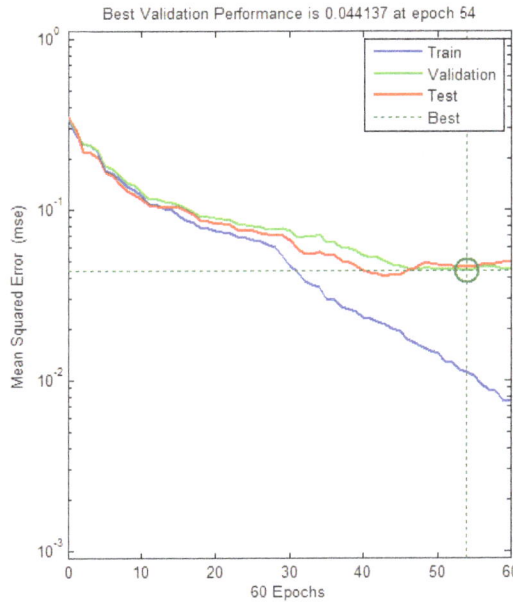

Figure 3. Network Performance

Table 4. Confusion matrix for blur classification results of blurred and noisy 1-D barcode images

Predicted Class		Target Class		
		Motion Blur	Defocus Blur	Combined Blur
	Motion Blur	199	10	0
	Defocus Blur	1	183	17
	Combined Blur	0	7	183

Table 5. Blur Classification results for blurred and noisy 1-D barcode images

Blur Type	Precision	Recall	Specificity
Motion Blur	95.2	99.5	97.5
Defocus Blur	91.0	91.5	99.5
Combined Blur	96.3	91.5	98.2
Accuracy			94.2

The training process is continued till the best validation performance achieved. The training stopped when the validation error increased for 6 iterations. The best validation performance is 0.0441, at epoch 54 as shown in Figure 3. The trained network model is used subsequently to classify the blur type from degraded images. Table 4 shows the confusion matrix of the proposed blur classification system. Table 5 shows the performance of the proposed model in terms of precision, sensitivity, specificity, and classification accuracy. The high values of these metrics show effectiveness of our model. The overall classification accuracy is 94.2 %.

Experiment 3: Performance analysis with blurred 2-D images

In this experiment, we have considered first 200 images from the 2-D barcode database. Afterward, the three different classes of blur (motion, defocus and combined blur) were artificially introduced to each barcode image with varying degree of parameters to construct the database of 600 blurred images (i.e., 200 images with each type of blur). Out of these 600 blurred images, 300, 120 and 180 images are utilized as training, validation and testing datasets respectively to design the neural model. The training process is continued till the best validation performance achieved. The training stopped when the validation error increased for 6 iterations. The best validation performance is 0.0067, at epoch 40 as shown in Figure 4. The trained network model is used subsequently to identify the blur type from degraded images. The overall blur classification results are presented as confusion matrix in Table 6. By this confusion matrix statistical quantities precision, recall, specificity and accuracy are calculated. Table 7 shows these metrics. The high values of these metrics show effectiveness of our model. The overall classification accuracy is achieved as 98.7 %.

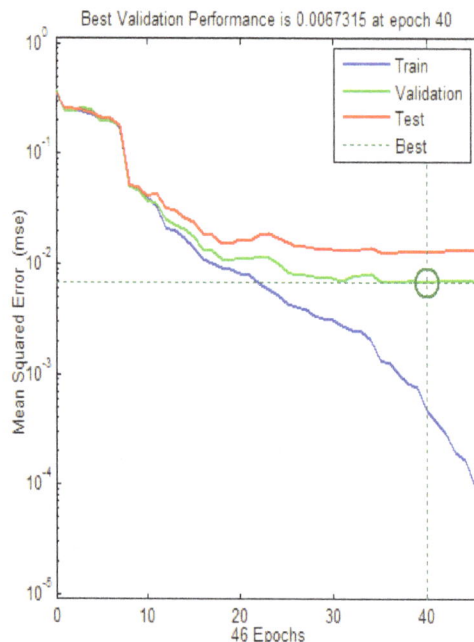

Figure 4. Network Performance

Table 6. Confusion matrix for blur classification results of blurred 2-D barcode images

Predicted Class	Target Class		
	Motion Blur	Defocus Blur	Combined Blur
Motion Blur	199	5	0
Defocus Blur	1	194	1
Combined Blur	0	1	199

Table 7. Blur Classification results for blurred 2-D barcode images

Blur Type	Precision	Recall	specificity
Motion Blur	97.5	99.5	98.7
Defocus Blur	99.0	97.0	99.5
Combined Blur	99.5	99.5	99.7
Accuracy			98.7

Experiment 4: Performance analysis with blurred and noisy 2-D images

To examine the efficiency of the proposed method in presence of noise, another neural model is created using 200 2-D barcode images. First, all the images are degraded different kinds of blur similar to Experiment 3. Then, Gaussian noise of 40dB BSNR is introduced to the blurred images to create a database of blurred noisy samples. The training process is continued till the best validation performance achieved. The training terminated when the validation error increased for 6 iterations. The best validation performance is 0.0188, at epoch 23 as shown in Figure 5. The trained network model is used subsequently to recognize the blur type from degraded images. The performance of model is presented in Table 8 and Table 9. The overall classification accuracy is 96.3 %.

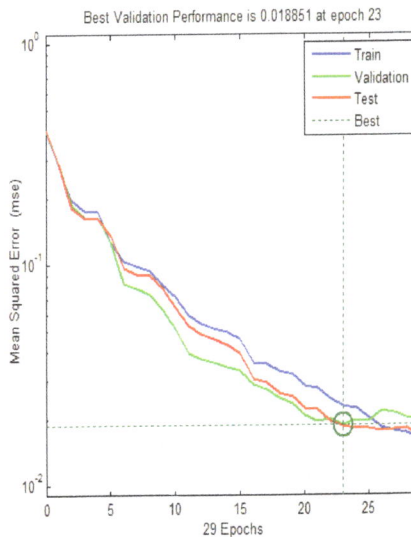

Figure 5. Network Performance

Table 8. Confusion matrix for blur classification results of blurred and noisy 2-D barcode images

Predicted Class	Target Class		
	Motion Blur	Defocus Blur	Combined Blur
Motion Blur	196	10	0
Defocus Blur	4	185	3
Combined Blur	0	5	197

Table 9. Blur Classification results for blurred and noisy 2-D barcode images

Blur Type	Precision	Recall	Specificity
Motion Blur	95.1	98.0	97.5
Defocus Blur	96.4	92.5	98.3
Combined Blur	97.5	98.5	98.7
Accuracy			96.3

In Table 10 results of proposed method are also compared with another method which uses statistical features for blur classification. It is obvious that proposed method gives better results in comparison to this method.

Table 10. A comparison of blur classification results

Experiment Data Sets	Statistical features based method [12]	Proposed method using curvelet energy features
1-D blurred images	96.7	98.2
1-D blurred and noisy images	95.0	94.2
2-D blurred images	98.0	98.7
2-D blurred and noisy images	95.2	96.3

7. Conclusion

In this paper, we have proposed a new blur classification scheme for barcode images taken by digital cameras. The scheme makes use of the ability of curvelet features to discriminate blur patterns appear in frequency spectrum of blurred images. This work identifies blur type, which can further help to choose the appropriate blur parameter estimation approach for blind restoration of barcode images. We have also show that in presence of low level noise, performance of blur classification system is precise. The limitation of this method is that presence of high level of noise causes disappearance of blur patterns in frequency spectrum, which cause the poor performance. Further work extension can be made to improve the performance of the classifier system with the high level of noise by incorporating the noise filtering techniques.

Acknowledgment

We highly appreciate College of Engineering and Technology, Mody University of Science & Technology, Lakashmangarh for providing facility to carry out this research work.

References

[1] Kato H, Tan KT. Pervasive 2D barcodes for camera phone applications. *IEEE Pervasive Computing.* 2007; 6(4): 76–85.
[2] Yang H, Alex C, Jiang X. Barization of low-quality barcode images captured by mobile phones using local window of adaptive location and size. *IEEE Trans. on Image Processing.* 2012; 21(1): 418–425.
[3] Joseph E, Pavlidis T. Bar code waveform recognition using peak locations. Pattern Analysis and Machine Intelligence. *IEEE Transactions on.* 1994; 16(6): 630-640.
[4] Selim E. Blind deconvolution of barcode signals. *Inverse Prob.* 2004; 20(1): 121– 135.
[5] Tiwari S, Shukla VP, Biradar SR, Singh AK. Review of motion blur estimation techniques. *J Image Graph.* 2013; 1(4): 176-184.
[6] Aizenberg I, Aizenberg N, Bregin T, Butakov C, Farberov E, Merzlyakov N, and Milukova O. Blur Recognition on the Neural Network based on Multi-Valued Neurons. *Journal of Image and Graphics.* 2000; 5: 127-130.
[7] Chong RM, Tanaka T. Image extrema analysis and blur detection with identification. *IEEE International Conference on Signal Image Technology and Internet Based Systems.* 2008: 320-326.
[8] Liu R, Li Z, Jia J. *Image partial blur detection and classification.* In Proc. CVPR. 2008: 23–28.
[9] Aizenberg I, Paliy DV, Zurada JM, Astola JT. Blur identification by multilayer neural network based on multivalued neurons. *IEEE Transactions on Neural Networks.* 2008; 19(5): 883-898.
[10] Yan R, Shao L. *Image Blur classification and parameter identification using two-stage deep belief networks.* British Machine Vision Conference (BMVC), Bristol, UK. 2013: 1-11.

[11] Tiwari S, Shukla VP, Sangappa Biradar AS. Texture Features based Blur Classification in Barcode Images. 2013.

[12] Tiwari S, Shukla VP, Biradar SR, Singh AK. Blur Classification Using Wavelet Transform and Feed Forward Neural Network. *International Journal of Modern Education and Computer Science (IJMECS).* 2014; 6(4): 16.

[13] Tiwari S, Shukla VP, Biradar SR, Singh AK. Blind Restoration of Motion Blurred Barcode Images using Ridgelet Transform and Radial Basis Function Neural Network. *Electronic Letters on Computer Vision and Image Analysis.* 2014; 13(3): 63-80.

[14] Tiwari S, Shukla VP, Biradar SR, Singh AK. Defocus Blur Parameter Estimation in Barcode Images using Fast Discrete Curvelet Transform. *International Journal of Tomography & Simulation™.* 2014; 27(3): 35-46.

[15] Tiwari S, Shukla VP, Biradar SR, Singh AK. Blur parameters identification for simultaneous defocus and motion blur. *CSI Transactions on ICT:* 1-12.

[16] Candes EJ, Donoho DL. Curvelets. Manuscript. http://www.stat.stanford.edu/~donoho/Reports/1998/curvelets.zip. 1999.

[17] Candes EJ, Demanet L, Donoho DL, Ying L. Fast discrete curvelet transform. *Technical report, CalTech.* 2005.

[18] Candès E, Demanet L, Donoho DL, Ying L. CurveLab-2.1. 2.2008-08-15. http://www.curvelet.org. 2008.

[19] Vasumathi G, P Subashini. Pixel Classification of SAR ice images using ANFIS-PSO Classifier. *Indonesian Journal of Electrical Engineering and Informatics (IJEEI)* 4.4. 2016.

[20] Azriyenni A, Mustafa MW, Sukma DY, Dame ME. 2014. Backpropagation Neural Network Modeling for Fault Location in Transmission Line 150 kV. *Indonesian Journal of Electrical Engineering and Informatics;* 2(1): 1-12.

[21] Wachenfeld S, Terlunen S, Jiang X. 1-D barcode image database. http://cvpr.uni-muenster.de/research/barcode. 2008.

[22] Szentandrási I, Dubská M, and Herout A. Fast detection and recognition of QR codes in high-resolution images. Graph@FIT, Brno Institute of Technology. 2012.

22

Design of Model Predictive Controller for Pasteurization Process

Tesfaye Alamirew[1], V. Balaji[2*], Nigus Gabbeye[3]
[1,3]Faculty of Chemical and Food Engineering, BIT, Bahir Dar University, Ethiopia
[2]Faculty of Electrical and Computer Engineering, BIT, Bahir Dar University, Ethiopia
e-mail: balajieee79@gmail.com

Abstract

This research paper is about developing a better type of controller, known as MPC (Model Predictive Control) for pasteurization process plant. MPC is an advanced control strategy that uses the internal dynamic model of the process and a history of past control moves and a combination of many different technologies to predict the future plant output.. The dynamics of the pasteurization process was estimated by using system identification from the experimental data. The quality of model structures like ARX, ARMAX, BJ and CT model structures was checked based on best fit with validation data, residual analysis and stability analysis. Auto-regressive with exogenous input (ARX322) model was chosen as a model structure of the pasteurization process dynamics and fits about 79.75% with validation data. Finally MPC control strategies were designed using ARX322 model structure.

Keywords: pasteurization process, MPC, ARX model, ARMAX model, BJ model, CT model

1. Introduction

In a modern world the economic and quality issues become more and more important, efficient control systems have become indispensable. Therefore the process industries require more reliable, accurate, robust, efficient and flexible control systems for the operation of process plant. In order to fulfill the above requirements there is a continuing need for research on improved forms of control. [1]

Control of temperature plays an important role in pasteurization plants. High temperature short time (HTST) is keeping milk or other food stuffs at 72 0C for 15 seconds in insulated holding tube. The pasteurization process consists of three stages like regeneration, heating and cooling sections. The crucial stage is heating process using heat exchanger to ensure unpasteurized product achieve desired pasteurization temperature before pass through holding tube and cooling sections. Prior to pasteurize milk sample, the equipment must have adequate controller to control the outlet temperature in order to maintain at standard value. [2]

The proportional integral (PI) and proportional integral derivative (PID) controllers are widely used in many industrial control systems because of its simple structure. These controllers are designed without process constraints only use mathematical expression based on error from a set point. In these circumstances, conventional controllers (PI and PID) are no longer to provide adequate and achievable control performance over the whole operating range. Thus designing a controller considering the process constraints and optimize the control performance is essential. [3]

Model Predictive Control also known as receding horizon control, is an advanced strategy for optimizing the performance of multivariable control systems. MPC generates control actions by optimizing an objective function repeatedly over a finite moving prediction horizon, within system constraints, and based on a model of the dynamic system to be controlled. [4]

2. Process Description

The plant PCT23, manufactured by Armfield (UK), is a laboratory version of a real industrial pasteurization process. It consists of a bench-mounted process unit to which is connected a dedicated control console. An interface card DT2811 is used for monitoring and controlling the process through a computer. [5]

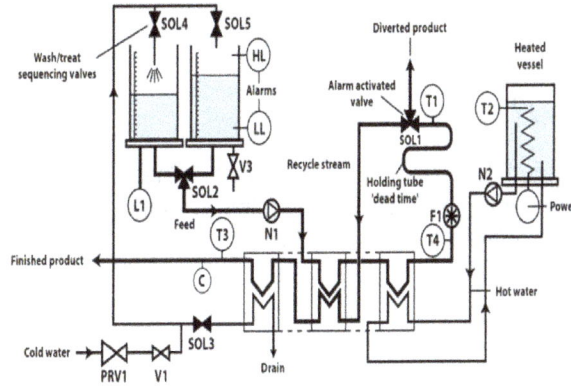

Figure 1. Process flow diagram of pasteurization plant

Here temperature T1 is the controlled variable and milk flow rate (N1), hot water flow rate (N2) and power are manipulated variables.

3. Experimental Setup
3.1. Input-Output Data
The input-output data was generated by introducing step input in milk flow rate, hot water flow rate and power input, then by recording pasteurization temperature response. The experiment was repeated two times for model estimation and validation purpose.

3.2. System identification
The input-output data was analyzed by the System Identification toolbox in MATLAB. The continuous and discrete model structures were tried to select the model structure that have best fit with validation. Then the selected model structure is tested for residual analysis and pole-zero analysis to check the model stability. The continuous time (CT) model, Auto-regressive with exogenous input (ARX) model structures, ARMAX (auto regressive with moving average and exogenous (or extra) input model, and state space model structures were tried get best model structure in terms of best fit with validation data and model stability for further controller design. [6]

Best fit is calculated as:

$$\text{Best fit } (\%) = \left(1 - \frac{(y - \hat{y})}{(y - \bar{y})}\right) * 100 \qquad (1)$$

where: y is validation data, $\hat{y}$ is estimated data and $\bar{y}$ is mean of validation data

After selection of best fit model structures model quality analysis like residual and pole zero location should be checked to select a nice and simple model for further controller design. The prediction error or residual is the key quantity.

It is defined as:

$$\epsilon(k) = y(k) - \hat{y}(k) \qquad (2)$$

The stability of a system can be easily inferred by examining the pole locations of the transfer function. [7].

3.3. Controller Design
Controllers are basically employed in a closed loop control system. Closed loop control system is one that automatically changes the output based on the difference between the feedback signals to the input signal. Controller is an element used to produce manipulated variable from error variable, for Control action. [8][9]

3.3.1. Model Predictive Control
The model predictive uses quadratic minimization problem defined as:

$$J = (y_{sp} - y)^T (y_{sp} - y)Q1 + \Delta u^T \Delta u Q2 \tag{3}$$

Subject to: input and output constraints of the system. Where y_{sp} is the set point, Q1 is output weight and Q2 is input weight. The size of this minimization function and weight matrixes are depend on prediction and control horizon. [10][11]

4. Results and Discussions
4.1. Model Structure selection
First step input was introduced at different time on milk flow rate, water flow rate and heater power to collect pasteurization temperature data with those three inputs. Two different experiments were done to collect the data for model estimation and validation purpose until it reaches to stability. The continuous time model fits 82.77% with the validation data better than the others. ARX422 (81.03% fit), ARMAX3202 (80.9% fit). The continuous time model doesn't mean a good model rather further analysis will be needed to select best model.

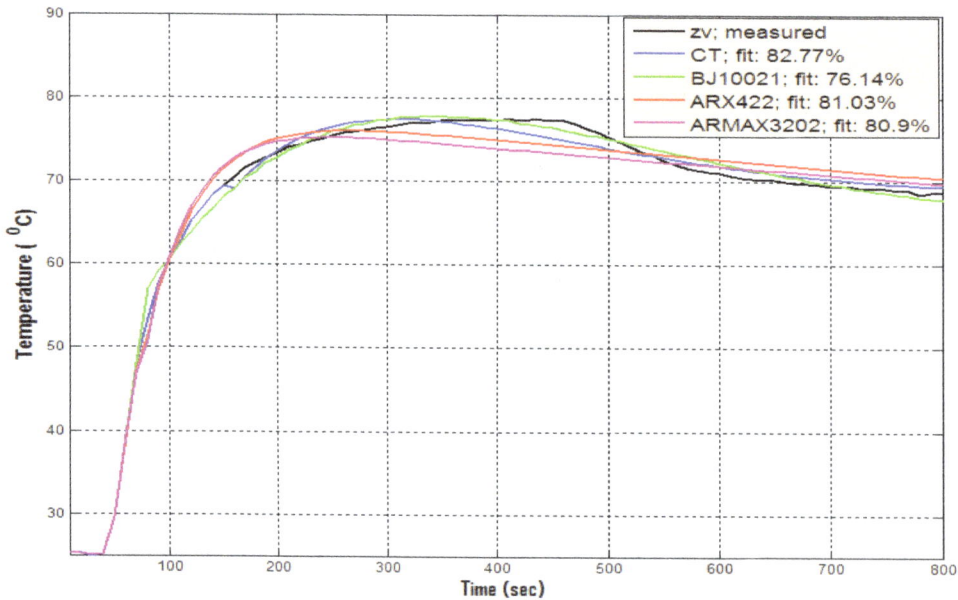

Figure 2. Percent fit of different model structures with validation data (zv).

4.2. Model Quality Analysis
4.2.1. Residual Analysis
For different model structures the auto corelation of residuals for the output (whitness test) and cross correlation of residuals with the input (independence test) were analyzed. From the graphs the horizontal scale is the number of lags, which is the time difference (in samples) between the signals at which the correlation is estimated. The upper and lower bounds on the plot represents the confidence interval of the corresponding estimates. Any fluctuations within the confidence interval are considered to be insignificant. A good model should have residuals uncorrelated with past inputs (independence test) and past outputs (whitness test). For poor models either auto and cross corelation residuals or two of residuals is out of the confidence region. In our case 99.9% confidence interval is taken. From Figure 3, the BJ10021 model is failed the analysis because both of auto and cross correlation residual analysis is out of the confidence region. The continuous time (CT) model also failed the analysis due to its auto correlation is out of the limit. ARMAX3202 and ARX422 models pass the residual analysis, but further analysis should be taken to select best model structure.

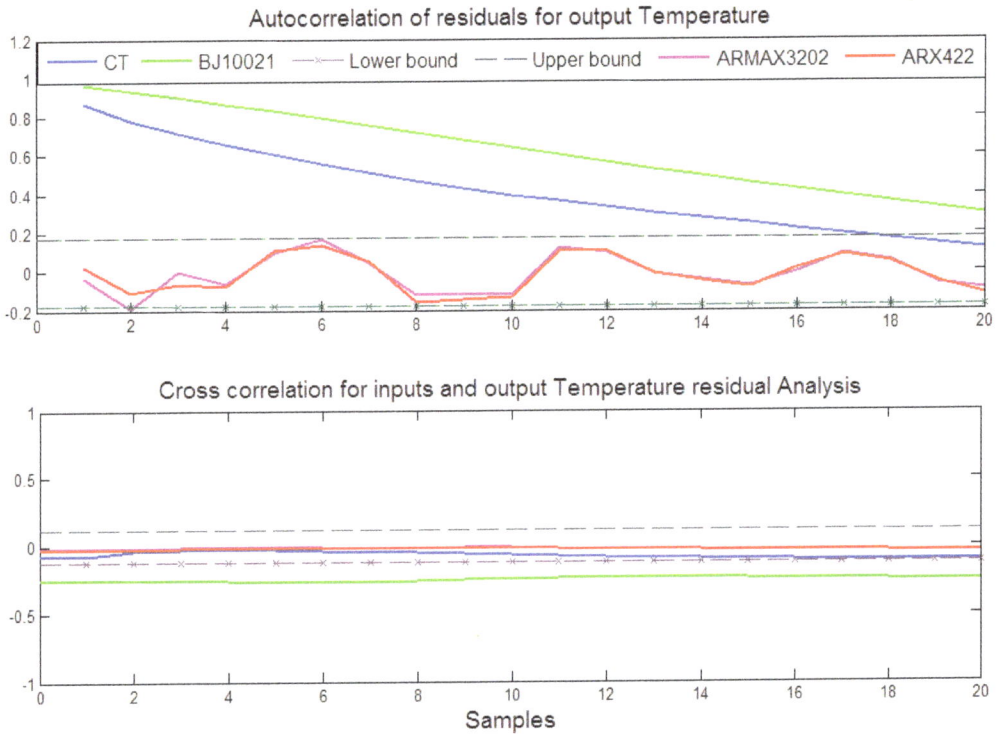

Figure 3. Residual analysis of model structures

4.2.2. Stability Analysis

The pole-zero location on the unit circle can tell as the stability of the process model. Poles are detrimental for the process stability. If all poles are inside a unit circle, the process model is stable. If one or more of its poles on unit circle, it is marginally stable. If one of its poles out of a unit circle, it is unstable.

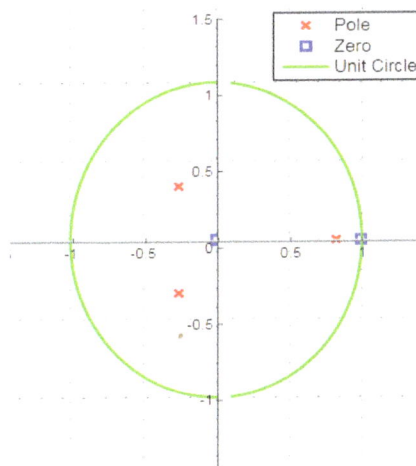

Figure 4. Pole-Zero location of ARX 422

When we see the pole-zero plot of the ARX422 model structure in Figure 4, it is stable because its entire pole is inside a unit circle in three of input - output relations. Some poles and zeros are lie on the same location. This means that we can cancel the numerator and denominator. Pole – zero cancelation may be an opportunity for model order reduction. Generally ARX422 model structure is stable and passes this analysis.

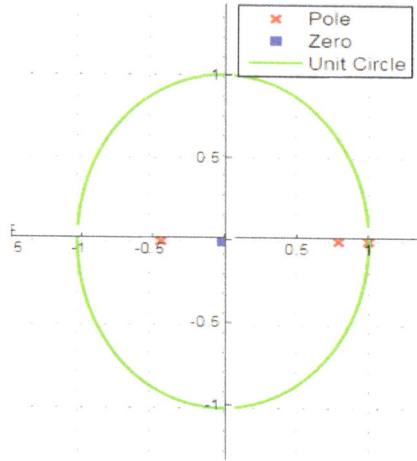

Figure 5. ARMAX3202 model structure pole-zero location

When we see Figure 5 one of its pole is on a unit circle, this means the process model is marginally stable. The process model is not selected because it has the chance to become unstable.

4.2.3. Model Reduction
When the model order reduced ARX gives slight decrease of fit.. When the ARX422 is reduced to ARX322 the final prediction error is 0.035. Therefore ARX322 can represent the model. Further reduction below this order deteriorates the fit percent with validation data. The reduced model also passes the model quality analysis. Therefore the ARX322 model represents the pasteurization process dynamics.

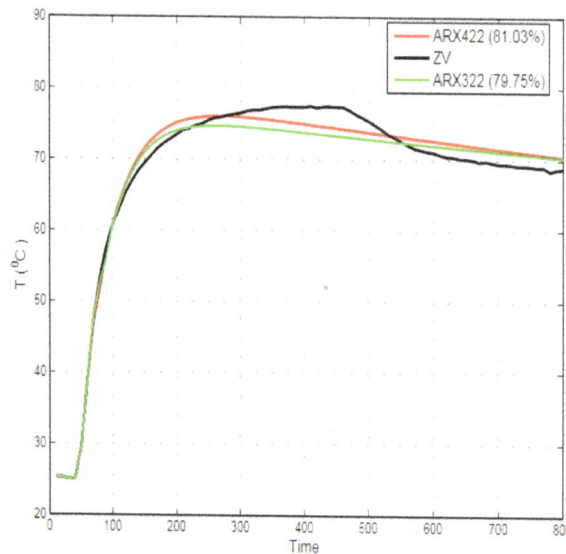

Figure 7. ARX422 and ARX322 model fit percent

The following equation is the converted discrete ARX model to continuous dynamic model for PCT23 pasteurization plant dynamics. The process has three inputs that come parallel and one output. The process dynamics is the third order process and described as shown below.

Continuous-time ARX322 model is represented as follows.

$$A(s)Y(t) = B(s)u(t) + C(s)e(t)$$

Therefore the estimated parameters look like this $A(s) = s^3 + 0.105s^2 + 0.0019s + 5.55 * 10^{-7}$

$$B1(s) = 0.0007s^2 + 5.22 * 10^{-5}s + 5.98 * 10^{-8}$$
$$B2(s) = 0.0001s^2 + 7.35 * 10^{-6}s - 3.75 * 10^{-8}$$
$$B3(s) = 0.0072s^2 + 0.0014s + 7.023 * 10^{-5}$$
$$C(s) = s^3 + 0.22s^2 + 0.019s + 0.00064$$

Input delays are 10, 80 and 30 respectively for the three inputs (milk flow rate, hot water flow rate and power input). The ARX322 model has Loss function of 0.0334283 and final prediction error (FPE) of 0.034922.

4.3. MPC Controller Design
Synthesis of MPC controller requires formation of an object's mathematical model, definition of cost function form and lengths of control and prediction horizons.

Lengthening of the prediction and control horizons elongates computation time, because it complicates optimization procedure. It also improves controller robustness. So there is a need to arbitrary and iteratively choose their lengths. The controller should be as robust to disturbances and as fast as possible.

The prediction horizon of the system should be large enough to cover the settling time.

The control horizon for different system should be different. It depends on the output signal of the system. In the most cases the control horizon should be large enough to get the reasonable stabilize output signal of the system.

Table1 : Model Predictive Controller parameter's values

Parameter	Value
Prediction Horizon length	200
Control Horizon length	95
Matrix of Weights for the Output Signal Q1	1 0 ... 0 / 0 1 ... ⋮ / ⋮ ⋮ ⋱ 0 / 0 ... 0 1
Matrix of Weights for the Input Signal Q2	0 0 ... 0 / 0 0 ... ⋮ / ⋮ ⋮ ⋱ 0 / 0 ... 0 0
Milk flow rate, mw [ml/min]	326
Hot water flow rate ,hw [ml/min)	0 < hw < 800
Power input Pi, [Kwh]	0 < Pi< 1.7

Offline simulation for pasteurization temperature response is shown in Figure 8. The output response is reasonably tracts the set point without any overshoot, but it is slow (sluggish) because of it has too long control horizon. Working in optimal condition by considering process constraints makes MPC controller has best performance.

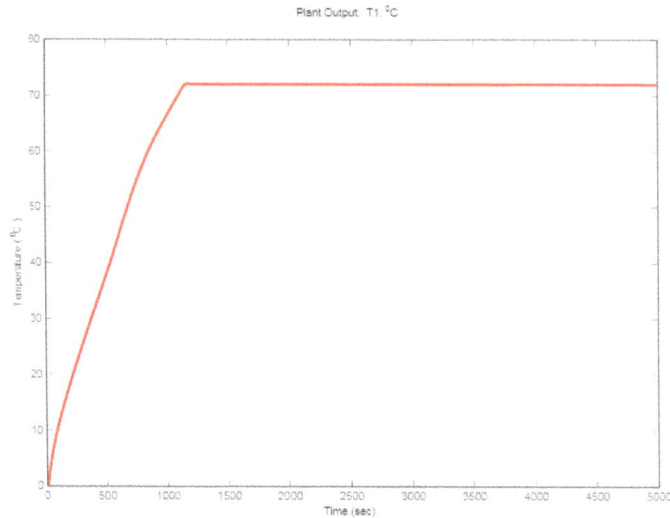

Figure 8. Pasteurization temperature response using MPC

5. Conclusions

Maintaining the temperature at a constant value is a critical issue in many of the Industries. MPC fulfills these types of difficulties by bringing the process variable to the desired set point as early as   possible. MPC controller is more suitable for complex process control like milk pasteurization processes. From the simulation results, the MPC controller removes overshoot, but the control action is sluggish. to track set point immediately. This controller performance may be best, if it is used for real time pasteurization process environment.

Acknowledgement

The authors wish to thank Faculty of chemical and food engineering process control laboratory workers for their technical support and also, Mr. Zewdu Tsegaye and Mr. Zelfikir Jemal for their comments and suggestions.

References

[1]   Diederich Hinrichsen, Anthony J Pritchard. Mathematical Systems Theory I. *Modeling, State Space Analysis, Stability and Robustness.* 2000; 1(6): 70.
[2]   Negiz A, Cinar A, Schlesser JE, Ramanauskas P, Armstrong DJ, Stroup W. *Automated control of high temperature short time pasteurization. Food Control* 7. 1996: 309–315.
[3]   Subhransu Padhee. Performance Evaluation of Different Conventional and intelligent. Sandviken, 80176 Gävle, Sweden.
[4]   Orukpe PE. Basics of Model Predictive Control. Imperial College, London. April 14, 2005.
[5]   ARMFIELD Process plant trainer (PCT23) Manual
[6]   Lennart Ljung. System Identification: Theory for the user. Second Edition, Linköping University, Sweden. Prentice Hall PTR. 1999.
[7]   Escobet, J Quevedo. Linear Model Identification Toolbox For Dynamic Systems. *IEEE.* 1998.
[8]   Astrom, Karl J, Hagglund, Tore. Advanced PID Control, Department of Automatic Control, Lund Institute of Technology, Lund University.
[9]   Yuvraj Bhusan Khare, Yaduvir Singh. PID control of Heat Exchanger System. Punjab, India, 2010.
[10] Sidharta Dash, Mihir Narayan Mohanty. Analysis of Outliers in System Identification using WMS algorithm. 2012.
[11] J Rossiter. Model-based predictive control: A Practical Approach. CRC Press LLC. 2003.
[12] MATLAB R2008b User Manual.
[13] F Yarman, BW Dickinson. *Autoregredon Estimation Using Final Prediction Error.* Proceedings of the IEEE. 1982; 70.
[14] Adriaan Van Den Bos. Parameter Estimation for Scientists and Engineers. Published by John Wiley & Sons, Inc, Hoboken, New Jersey, 2007.

[15] Garcia CE, Prett DM, Morari M. Model Predictive Control: Theory and Practice - a Survey. Automatica. 1989; 25(3): 335-348.

[16] Antonio Balsemin. Applications Oriented Input design for MPC: An analysis of a Quadruple water tank process. KTH Electrical Engineering degree project, Stockholm, Sweden. August 2012.

[17] Bequette, B Wayne. Process Control: Modeling, Design and Simulation. Prentice Hall PTR. 2002: 487-511.

[18] EC Kerrigan, JM Maciejowski. *Soft constraints and exact penalty functions in Model predictive control.* in Proc. UKACC International Conference. 2000.

23

A Lyapunov Based Approach to Enchance Wind Turbine Stability

A. Bennouk*, A. Nejmi, M. Ramzi
Mathematics and physics laboratory
Faculty of Sciences and Technics, Beni Mellal
Sultan Moulay Slimane University, Beni Mellal, Morocco
e-mail: bennouk.anasse@gmail.com

Abstract

This paper introduces a nonlinear control of a wind turbine based on a Double Feed Induction Generator. The Rotor Side converter is controlled by using field oriented control and Backstepping strategy to enhance the dynamic stability response. The Grid Side converter is controlled by a sliding mode. These methods aim to increase dynamic system stability for variable wind speed. Hence, The Doubly Fed Induction Generator (DFIG) is studied in order to illustrate its behavior in case of severe disturbance, and its dynamic response in grid connected mode for variable speed wind operation. The model is presented and simulated under Matlab/ Simulink.

Keywords: Double Fed Induction Generator, wind turbine, Rotor Side Converter, Lyapunov, Grid Side Converter

1. Introduction

Today, the number of wind turbines connected to the grid is steadily increasing. Variable speed wind turbines outperform constant speed turbines in aerodynamic efficiency while also reducing stress on the mechanical apparatus so as to make it more productive and cost-effective [1].

Double Fed induction generator (DFIGs) is becoming more used in modern wind power generation systems due to their variable speed operation, four quadrant active and reactive power regulation, low converter cost, and reduced power losses compared with Permanent Magnet Synchronous generators (PMSG).

Classic control of grid connected to DFIG is usually based on VOC (Voltage Oriented Control) or FOC (Flux Oriented Control), it decouples the d and q rotor currents in the synchronous frame. Control of instantaneous active and reactive power is then achieved by regulating the decoupled rotor currents, using proportional–integral (PI) parameters and accurate machine parameters. The weakness is shown when the machine's converters operate beyond their linear limits [1].

Field Oriented Control (FOC) is improved in this paper by the use of Backstepping strategy to control the Rotor Side inverter; this method has scored positive performances. This kind of control is generally associated with Lyapunov functions in order to increase system stability against variable wind fluctuations.

Moreover, Variable structure control or Sliding Mode Control (SMC) strategy is an effective and a high-frequency switching control for nonlinear systems showing uncertainties. It features simple implementation disturbance rejection, strong robustness and fast responses. This method will be used in the Grid Side inverter control so as to enhance Udc stability and then reach the grid parameters in term of frequency, Total Harmonics Distortion (THD) and voltage unbalance [2].

This paper is organized as follows: the second section is due to deal with a wind energy modeling under Matlab/Simulink. The third section illustrates the Field Oriented Control and Backstepping linearization control for the rotor side converter. The fourth section introduces a SMC destined to control of the Grid Side Converter performances. Finally, a simulation with its inherent results will be presented.

2. Wind Turbine Modeling
This section presents WECS based on a DFIG connected to the electric grid

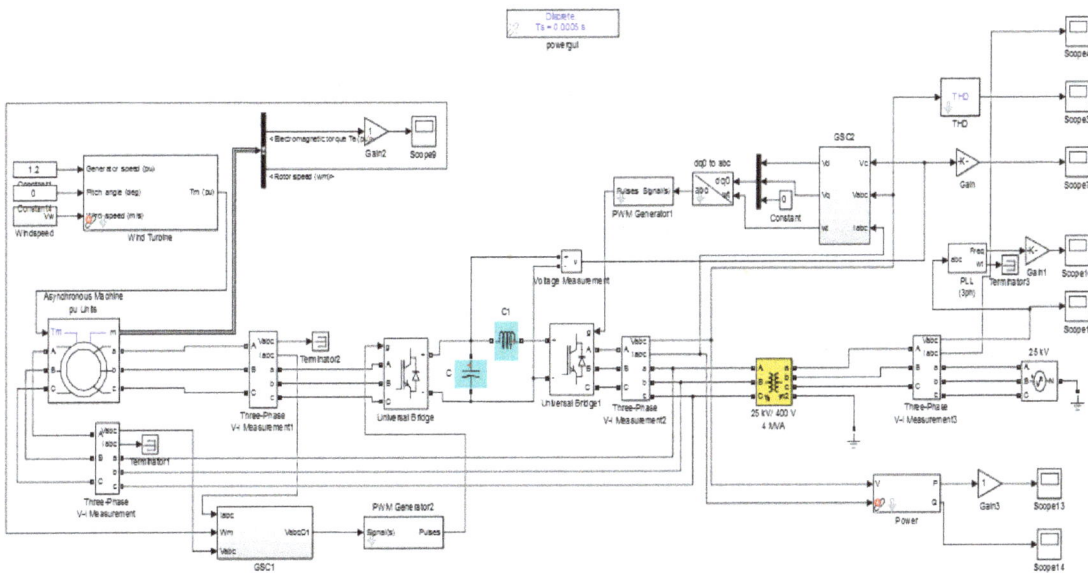

Figure 1. Wind turbine based on DFIG

Figure 1 shows the scheme of electrical energy's generation from the wind power on the basis of DFIG. As it is demonstrated, the stator is connected directly to the grid whereas the rotor is connected to the grid via a Back-to-back converter [2].

2.1. Wind Speed Model
Wind speed model contains four components [1]:

$$V_w(t) = V_b(t) + V_r(t) + V_n(t) + V_g(t) \tag{1}$$

where V_b is the base wind component (constant), V_r is the ramp wind component, V_n is the base noise wind component and V_g is the gust wind, V_g is set to zero during simulation, all of them in m/s. The model implementation of the wind speed in Matlab/Simulink is presented in Figure 2:

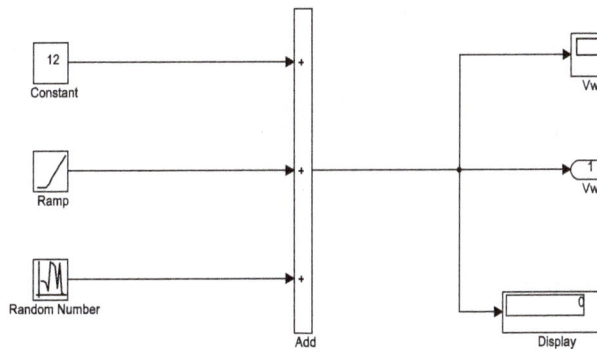

Figure 2. Wind speed model

The kinetic energy in this level is given by:

$$P_w = \frac{1}{2}mv^2 = \frac{1}{2}\rho S v^3 \qquad (2)$$

where m is the air mass, ρ is the air density, v is the wind speed and S is the covered surface of the turbine.

2.2. Wind Turbine Modeling

Wind turbine is applied to convert the wind energy into mechanical torque. The mechanical torque of the turbine can be calculated from mechanical power at the turbine extracted from wind power. Then, the power coefficient of the turbine (Cp) is used. The power coefficient is function of pitch angle (β) and tip speed (λ). The power coefficient maximum of (C_p) is known as the limit of Betz [2].

The power coefficient is given by:

$$C_p = C_1 \left(\frac{C_2}{\lambda i} - C_3\beta - C_4\right) e^{-\frac{C5}{\lambda i}} + C_6\lambda \qquad (3)$$

where

$$\frac{1}{\lambda_i} = \frac{1}{\lambda+0.08\beta} - \frac{0.035}{\beta_3+1} \quad \text{and} \quad \lambda = \frac{\omega_m.R}{v}$$

C_1, C_2, C_3, C_4, C_5 and C_6 are constants given by the turbine constructor (C_1 = 0.516, C_2 = 116, C_3 = 0.4, C_4 = 5.1, C_5= 21 and C_6 =0.0068), R is the rotor radius and ω_m is the rotor speed generator. The power coefficient is a nonlinear function of the tip speed ratio λ and the blade pitch angle β (in degrees).If the swept area of the blade and the air density are constant, the value of C_p is a function of λ and it is maximum at the particular λ optimum. Hence, to fully utilize the wind energy, λ should be maintained at λ_{opt}, which is determined from the blade design, C_p is defined as [3]:

$$C_p = \frac{Pm}{Pw} \leq 1$$

and

$$P_m = \frac{1}{2}C_p\rho S v^3 \qquad (4)$$

where P_m is the mechanical output power of the turbine.

The Rotor Side converter is used to extract the MPPT (Maximum Power Point Tracking), in this purpose the DFIG angular speed reference is calculated permanently with an approach to follow the wind speed fluctuation, the wind turbine control generates the DFIG speed reference signal, performed by the RSC controller in DFIG control level. This reference signal is determined from the predefined characteristic P-ω, based on filtered measured generator speed. Whereas the Grid Side Converter is used to control the DC link voltage and guarantees unity power factor in the rotor branch. The transmission of the reactive power from DFIG to the grid is thus only through the stator, a conventional vector with Lyapunov approach will be used in this command algorithm to enhance system stability.

3. Rotor Side Converter control

The standard vector control is used to control the RSC. In order to ensure stability when applying this control strategy mechanism, the stator flux magnitude must remain constant. To this purpose, Lyapunov's approach is used in order to enhance the system's stability [2].

The electrical energy conversion is described by the following equation according to a d-q frame :

$$V_{sd} = R_s i_{sd} + \frac{d\varphi sd}{dt} - \omega_s \varphi_{sq} \tag{5}$$

$$V_{sq} = R_s i_{sq} + \frac{d\varphi sq}{dt} + \omega_s \varphi_{sd}$$

$$V_{rd} = R_r i_{rd} + \frac{d\varphi rd}{dt} - \omega_r \varphi_{rq}$$

$$V_{rq} = R_r i_{rq} + \frac{d\varphi sq}{dt} + \omega_r \varphi_{rd}$$

and

$$\varphi_{sd} = L_s i_{sd} + L_m i_{rd} \tag{6}$$

$$\varphi_{sq} = L_s i_{sq} + L_m i_{rq}$$

$$\varphi_{rd} = L_r i_{rd} + L_m i_{sd}$$

$$\varphi_{rq} = L_r i_{rq} + L_m i_{sq}$$

Electromagnetic torque is expressed by:

$$\text{Te} = \frac{P}{\Omega} = \frac{P.p}{\omega} = p(\varphi_{sd} i_{rq} - \varphi_{sq} i_{rd}) \tag{7}$$

V is the voltage, R_s and R_r are respectively the stator and rotor resistance. L_s, L_r and L_m are stator, rotor and mutual inductance between stator and rotor θ_s, and θ_r present angles of stator and rotor frames.

The aim of this approach is to catch the MPPT and to stabilize active power through regulating the electromagnetic torque depending basically on i_{rq} and i_{rd}, the stator flux is, thus, oriented to d axis in order to simplify the equations [5], [6]:

$$V_{sd} = 0 \text{ and } V_{sq} = \omega_s \varphi_{sq} \tag{8}$$

$$V_{rd} = R_r i_{rd} + \frac{d\varphi_{rd}}{dt} - \omega_r \varphi_{rq}$$

$$V_{rq} = R_r i_{rq} + \frac{d\varphi sq}{dt} + \omega_r \varphi_{rd}$$

The new system inputs are respectively d-q rotor current and angular speed expressed by the following equations:

$$\frac{dird}{dt} - \frac{1}{Lr}vrd - \frac{Rr}{Lr}ird + \omega rirq\left(\frac{Ls.Lr - M^2}{Ls.Lr}\right) \tag{9}$$

$$\frac{d\varphi rq}{dt} = Vrq - Rsird - \omega r\varphi rd$$

$$\frac{dirq}{dt} = -\frac{1}{Lr}vrq + \left(\frac{M^2}{LsLr} - \omega r\right)ird - \frac{M\omega r}{LrLs\omega s}Vsq - \frac{Rr}{Lr}irq$$

$$\frac{d\omega}{dt} = \frac{Cr}{J} - \frac{B}{J}\omega - \frac{K\varphi sdirq}{J}$$

It is clear from the dynamic model above, the nonlinearity, because of the coupling between the d-q rotor currents and the speed, a variable change is initiated so as to introduce the Lyapunov function:

$$y_1 = \omega_c - \omega \tag{10}$$

$$y_2 = i_{rdréf} - i_{rd}$$

$$y_3 = i_{rqréf} - i_{rq}$$

ω is the angular speed, ω_c is the reference of ω, i_{rd} and i_{rq} are respectively the d and q axis rotor current, , $i_{rdréf}$ and $i_{rqréf}$ are their references.
Lyupunov function is chosen as :

$$V1 = \frac{1}{2}y_1{}^2 \tag{11}$$

$$\dot{V}_1 = y_1\dot{y}_1 = y_1(\dot{\omega}_c - Kp\varphi_{sd}i_{rq} + \frac{B}{J}\omega + \frac{Cr}{J}) \tag{12}$$

The derivative of the complete Lyupunov function is negative defined, if the quantities between parentheses in equation (12) are equal to zero. i$_{qref}$ is given by:

$$i_{rqref} = \frac{1}{Kp\varphi sd}(Cr + B.\omega + 2.J.K.y_1) \tag{13}$$

$$\dot{y2} = \imath d\dot{r}ef - \dot{\imath_{rd}} - Ky_2 + Ky_2$$

and

$$\dot{y3} = \imath q\dot{r}ef - \dot{\imath_{rq}} - Ky_3 + K3$$

V 2 and V3 are defined respectively as :

$$V2 = \frac{1}{2}y_1{}^2 \ and \ V3 = \frac{1}{2}y_3{}^2 \tag{14}$$

Lyapunov function is defined as:

$$V4 = \frac{1}{2}(y_1{}^2 + y_2{}^2 + y_3{}^2) \tag{15}$$

$$\dot{V}4 = -Ky_1{}^2 - Ky_2{}^2 - Ky_3{}^2 + y_2(\imath d\dot{r}ef - \dot{\imath_{rd}} + Ky2) + y_3(\imath q\dot{r}ef - \dot{\imath_{rq}} + Ky3)$$

In order to fulfill Lyapunov's conditions, we integrate the following equations [2]:

$$A1 = y_2(\imath d\dot{r}ef - \dot{\imath_{rd}} + Ky_2) = 0 \tag{16}$$

and

$$A2 = y_3(\imath q\dot{r}ef - \dot{\imath_{rq}} + Ky_3) = 0 \tag{17}$$

We conclude after calculating direct and quadrature voltage that :

$$V_{rd} = -KL_ry_2 - R_ri_{rd} + \omega_r\left(L_sL_r - \frac{M^2}{LsLr}\right)i_{rq} \tag{18}$$

$$V_{rq} = -KLry_3 + \left(\frac{M^2}{L_sL_r} - L_r\omega_r\right)i_{rd} + \frac{M\omega r}{L_s\omega s}V_{sq} + R_r i_{rq}$$

4. Grid Side Converter control

When the grid voltage changes suddenly, a control approach is adopted in the GSC to keep the DC voltage constant and to assure a zero rotor reactive power [7], [8], [9].
First, the system equations are defined as follows:

$$Ud = -L\frac{did}{dt} - Ri_d + \omega Li_q + vd \tag{19}$$

$$Uq = -L\frac{diq}{dt} - Ri_q - \omega Li_d + vq$$

$$C\frac{d}{dt}vdc = i_{dc} = i_L - i_d$$

U_d, U_q :d-q components of converter voltage
v_d, v_q : d-q grid voltage components
i_d, i_q :d-q current components

We suppose that the grid voltage is aligned to the d-axis. We may define the state system as [10]:

$$\begin{bmatrix} \dot{\imath d} \\ \imath q \\ vdc \end{bmatrix} = \begin{bmatrix} -R/L & w \\ w & -\frac{R}{L} \\ -1/C & 0 \end{bmatrix} \begin{bmatrix} id \\ iq \end{bmatrix} + \begin{bmatrix} -\frac{Ud}{L} + \frac{vd}{L} \\ -1/LUq \\ iL \end{bmatrix} \tag{20}$$

$$y1 = id \tag{21}$$
$$y2 = vdc + iq - iL$$
$$S1 = y1ref - y1 \tag{22}$$
$$S_2 = y2ref - y2$$

In order to keep the system stability, the sliding system should be:

$$S_2.\dot{S}_1 \leq 0 \ and \ S_2.\dot{S}_2 \leq 0$$

Then we deduce the Ud and Uq :

$$U_d = Ksign\ (S_1) - Ri_q + L\omega i_d + v_d \tag{23}$$

$$U_q = Ksign(S_2) + i_d L\left(\omega - \frac{1}{C}\right) - Ri_q + (i_L - \dot{i}_L)$$

5. Simulation and results

The test wind profile with full field turbulence is generated through the use of Wind turbine block as presented in Figure 1. This block shows the hub height wind speed profile. In general, wind speed consists of two components, mean wind speed and turbulence component. The simulation is realized in order to illustrate Grid parameters stability, Grid voltage unbalance, THD and Frequency.

The wind turbine and DFIG parameters are illustrated in Table 1 [11]:

Table 1. DFIG parameters

Parameter	Symbol	Value
Active Power	P	2MW
Stator resistance	Rs	0.001518 Ω
Stator inductance	Ls	0.059906 H
Rotor inductance	Lr	0.082060 H
Rotor resistance	Rr	0.002087 Ω
Mutual inductance	Lm	2.4 H
Pole pairs	N	2
Inertia	J	17.23Kg.m²
Gear box	n_g	5.065

Simulation and results are presented first with a constant wind profile, then during a variable wind speed to illustrate the used approach robustness based on Lyapunov theory:

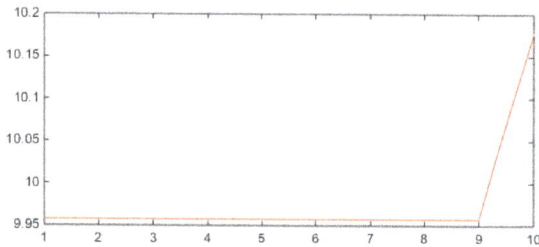

Figure 3. Wind speed

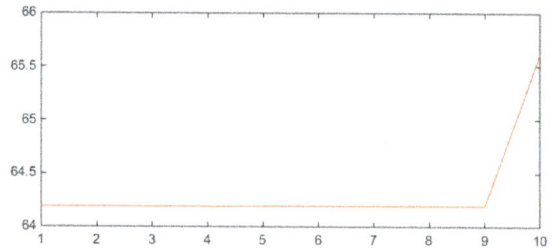

Figure 4. Rotor angular speed

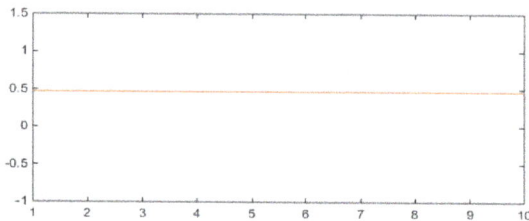

Figure 5. Power Coefficient (Cp)

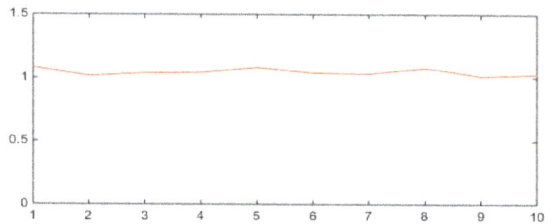

Figure 6. Vdc

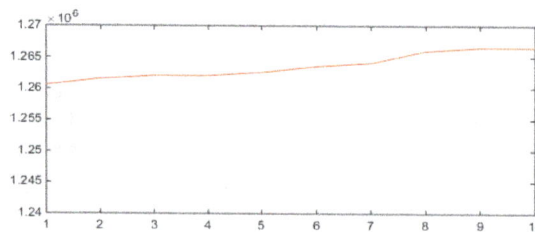

Figure 7. Active Power

It's obvious that for a constant wind speed,the wind power captured and deliverd to the grid had the same shape as the wind, active power is kept constant and equal to 1.26MW for a wind speed of 9.9m/s, the MPPT approach is well achieved and the power coeffiscient is maintained at 0.47. The DFIG speed is the image of the wind speed, it is following properly its reference. Lyapunov apporach used to control GSC shows also good performances and the Udc voltage was equal to its reference .

We present here simulations result following variable wind speed :

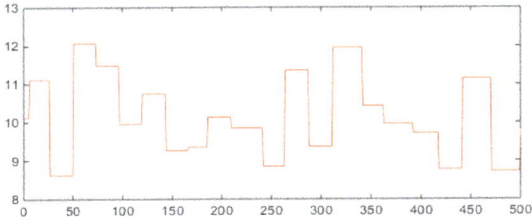

Figure 8. Variable wind speed

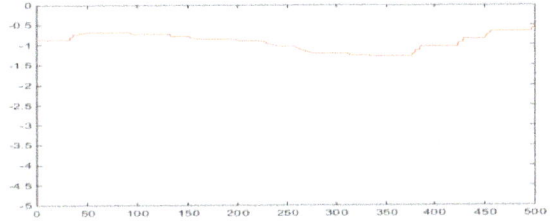

Figure 9. Electromagnetic torque

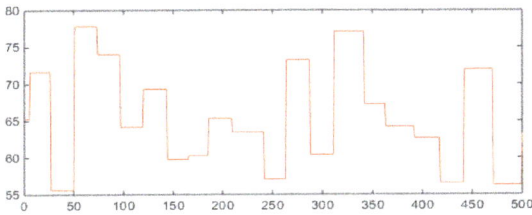

Figure 10. Rotor angular speed

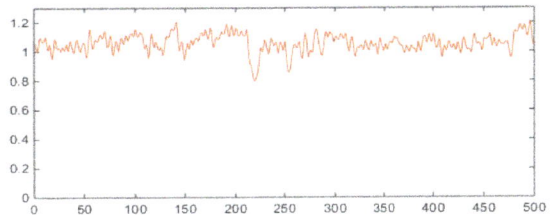

Figure 11. Vdc voltage

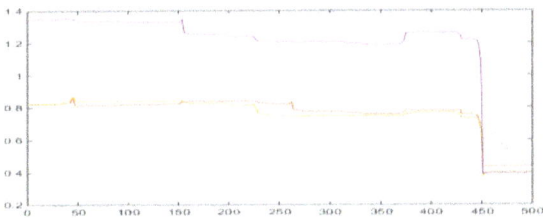

Figure 12. Total Harmonics Distorsion (THD)

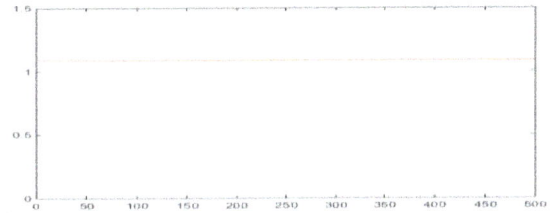

Figure 13. Frequency

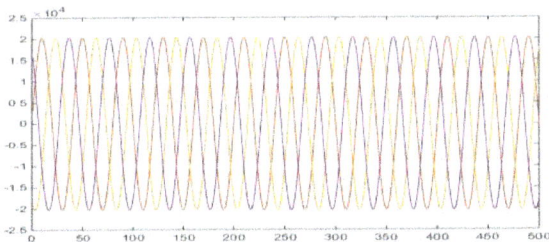

Figure 14. Grid voltage

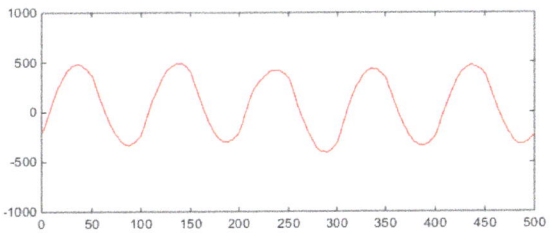

Figure 15. Grid Current

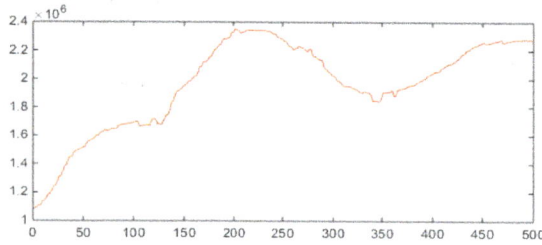

Figure 16. Active power

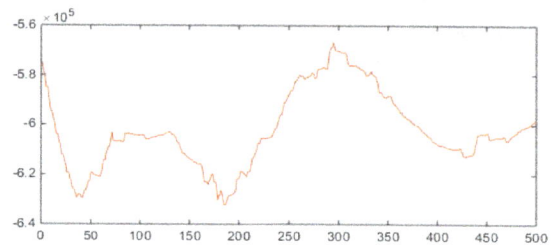

Figure 17. Reactive power

The dynamic responses of the DFIG generation system under an intermittent wind action are shown in Figure 8. It is clear that when a mean wind speed changes, the electromagnetic torque varies. We may deduce then that when the wind speed increases 11m/s, the electromagnetic field is set at 1pu. Consequently, the delivered power is more or equal to 2MW. Conversely, when the wind speed is inferior, the electromagnetic and active power generated decreases proportionally. The DFIG speed is the image of the wind speed, it is following properly its reference

The DC link voltage also varies according to the wind speed fluctuations, but the variation is not obvious as shown in Figure 11. The DC voltage is generally kept at its reference.
The Total Harmonic Distortion presented in Figure 12 had demonstrated good performances and did not exceed 5%, the assigned threshold of national grid code. The frequency, as shown in Figure 13, did not also exceed the threshold of 1.2 pu. The grid side voltage had shown also good performances apart from some a slight distortion caused by dynamic response and wind speed fluctuations. The balance voltage remained inferior to 1%, which is the fixed threshold of National Grid Code [2].
The active power variations illustrated in Figure.16 depend and follow the wind speed variation which has demonstrated also good performances. The active power reaches its highest values when the wind speed exceeds11m/s.
The reactive power had shown a good performance as presented in Figure.17 except in 200s, when we observe that the reactive power exceeds 600 Kvar following a quick fluctuation of Udc link voltage.

6. Conclusion

With the growing level of penetration scored by wind-origin power production into the general power system, many national codes have been applied so as to ensure stability to the electric grid. Many simulations and complete models are needed to be established prior to any connection between wind farm and electric grid [2].
A nonlinear control of variable speed WT is, thus, proposed. The main aim here is to maximize the energy capture from the wind while reducing Grid parameters deviations.
Also, and with the aim of illustrating from one side the system stability and to what extend it matches the national grid code, we have identified and presented the THD, frequency and voltage unbalance in the Point of Common Coupling (PCC), as these parameters had shown when they were under their limit during the simulation time.
Stability had been improved by using FOC and Lyupunov conditions in the Generator Side Converter. In the Grid Side Converter, SMC control was applied. The different PID controllers had been turned so as to obtain good performances. The use of SMC associated to Lyapunov had also demonstrated good performances in terms of Grid parameters stability.

References
[1] Hu J, Nian H, Hu B, He Y, Zhu ZQ. Direct Active and Reactive Power Regulation of DFIG Using Sliding –Mode Control. *IEEE*. 2010; 25(4): 1028-1039.
[2] Bennouk A, Nejmi A, Benamou A, Ramzi M. Backstepping and MIMO approachs to control a wind turbine based on DFIG. *International Journal of Emerging Technology and Advanced Engineering*. 2016; 6(3): 12-17.

[3]   Mullane A, Lightbody G, Yacamini R. Adaptive Control of Variable Speed Wind Turbines. *Rev. Energ.: Ren. Power Engineering*. 2001; 101-110.

[4]   Bossouf B, Karim M, Lagrioui A and Taoussi M. Backstepping Control of DFIG Generators for Wide-Range Variable-speed Wind Turbines. *Int. J. Automation and Control*. 2014; 8(2): 122-140.

[5]   Song Y.D, Dhinakaran B, Bao X.Y. Variable speed control of wind turbines using nonlinear and adaptive algorithms. *Journal of Wind Engineering and Industrial Aerodynamics*. 2000; 293-308.

[6]   Beltran B, Ahmed-Ali T, Benbouzid M. Sliding Mode Power Control of Variable-Speed Wind Energy Conversion Systems. *IEEE Transaction Energy Conversion*. 2008; 23 (8): 551-558.

[7]   Khemiri N, Kheder A, Faouzi M. An Adaptive Nonlinear Backstepping Control of DFIG Driven by Wind Turbine. WSEAS Transactions on Environment and Development. 2012; 8(2): 60-71.

[8]   Beltran B, Ali T, Benbouzid M. Sliding Mode Power Control of Variable-Speed Wind Energy Conversion Systems. *IEEE*. 23(2): 551-558.

[9]   Rajendran S, Jena D. Backstepping Sliding Mode Control of a Variable Speed Wind Turbine For Power Optimization. *Journal of Modern Power Systems and Clean Energy*. 2015; 3(3):402–410

[10]   Sung-Hun L, jun Joo Y, Juhoon Back, Jin-Heon S, Ick C. Sliding Mode Controller for Torque and Pitch Control of PMSG Wind Power Systems. *Journal of Power Electronics*. 2011; 11(3): 342-349.

[11]   Li S, Haskew T.A, Williams K.A, Swatloski R.P. Control of DFIG Wind Turbine With Direct-Current Vector Control Configuration. *IEEE*. 2012; 3(1): 1-11.

24

Application of SVC on IEEE 6 Bus System for Optimization of Voltage Stability

Sita Singh*[1], Jitendra Hanumant[2], Ashutosh Kashiv[3]
Electrical & Electronics Engineering Department, Oriental University
Opposite Revati Range Gate no. 1, Sanwer Road, Jakhya, Indore, MP, India 453555
e-mail: sitasigh200711@gmail.com[1], jitendrahanumant9@gmail.com[2], ashutosh.kashiv@gmail.com[3]

Abstract

The problem of voltage or current unbalance is gaining more attention recently with the increasing awareness on power quality. Excessive unbalance among the phase voltages or currents of a three phase power system has always been a concern to expert power engineers. The study of shunt connected FACTS devices is an associated field with the problem of reactive power compensation related problems in today's world. In this study an IEEE-6 bus system has been studied & utilized in order to study the shunt operation of FACTS controller to optimize the voltage stability

Keywords: *voltage stability, voltage collapse, Newton Raphson for load flow, SVC, IEEE – 6 bus system*

1. Introduction

At the present time, power systems are forced to operate at almost full capacity. More and more often, generation patterns result in heavy flows that tend to incur greater losses as well as threatening stability and security of the system. This ultimately creates undesirably increased risk of power outages of different levels of severity [1]. A traditional alternative to reinforce the power network consists of upgrading the electrical transmission system infrastructure through the addition of new transmission lines, substations, and associated equipment. However, the processes to allow, locate, and create new transmission line has become tricky, costly, time taking and numerous times even controversial [7].

On the other hand, FACTS device, which can provide direct and flexible control of power transfer and are very helpful in the operation of power network. When it is been discussed about the power system performance and the power system stability we can enhance by using FACTS device [8-9]. Static VAR compensator (SVC) is one of the most effective measure device for enhancing the power stability and power transfer capability of transmission network, in this SVC it should be properly installed in the system with uniform parameter setting. The some factors considering for optimal installation and the optimal parameter of SVC, in which we improve Stability margin, power loss decline, power collapse avoidance and power transmission capability enhancement [15, 12, 20]. This study deals with the objective to optimize power system voltage stability .This is achieved through IEEE- 6 Bus system, using Newton Raphson load flow analysis and then by placing the SVC on the weakest bus to attain the maximum possible voltage stability.

2. Voltage Stability & Collapse

A power system is claimed to be voltage stable if it is ready to maintain voltages similar to the steady values once subjected to small disturbances. At any instant of your time, the ability system operative condition should be in stable limits, summit completely different operational criteria; furthermore it ought to even be secure within the event of any credible incident [1]. Voltage instability issues disturbances during installation of network wherever the voltage magnitude becomes uncontrollable and eventually ensuing into a collapse of voltage magnitude. The voltage decline is usually monotonous within the starting of the collapse and sophisticated to note. Voltage decline increase typically marks the top of the collapse. It\'s troublesome to differentiate this development from transient stability whereas voltages can also decrease in a manner just like voltage collapse [2, 17]. Post-disturbance analysis could solely be in those cases reveal the actual cause throughout the last twenty years there are one or many giant voltage collapses virtually per annum somewhere round the world. The explanation behind it is terribly clear, those is that the accumulated variety of interconnections and a better degree of utilization of the ability system and alter in load characteristics. Nearly every type of contingencies and even slow-developing load will increase might cause a voltage stability drawback [11]. The duration for the course of events that change into a collapse varies from seconds to many tens of minutes. This makes voltage collapse troublesome to investigate since there are several other phenomena that act together around this point. Necessary factors that cause interaction throughout a voltage decline are: generation limitation, behaviour of on-load tap changers, and load characteristic. A motivating purpose is that several researchers discard voltage magnitude as an appropriate indicator for the proximity to voltage collapse, though this is often indeed the amount that collapses. One question that has been mentioned is whether or not voltage stability could be a static or dynamic method. These days it is widely accepted as a dynamic development however abundant analysis is performed victimises static models [3-5].

3. Proposed Work

SVC is one of the most effective devices for enhancing the voltage magnitude stability known. An IEEE-6 bus system is studied using the load flow equations through MATLAB programming. Load flow solutions are being carried out through Newton-Raphson Method. First this is done without the SVC for the load change on the weakest bus known and then the collapse point for the same is been figured out. Then with the collapse point reactive power at load side is increased by 5%, 10% and so on in steps of 5% upto 30% and have performed power flow analysis for each, a report for the same has been stored for comparative analysis. The results of the original system & the system with SVC implemented are hence found out to be almost same and hence the system is voltage stable with SVC even on maximum load on the weakest bus. Power flow analysis is been performed using Newton Raphson Algorithm which is explained as in next section.

4. Algorithm Used

Algorithm: Newton- Raphson for load flow analysis [13, 15]

---

1. Input: Read Bus Data and line Data
2. Assume a suitable solution for all buses except for the slack bus. Let Vp =1 + j0 for p = 1,2....n, p≠s, Vs = a + j0. Where s = slack bus
3. Set convergence criteria = ε i.e., if the largest of the residues exceeds ε the process is repeated, otherwise is terminated.
4. Set Iteration count K = 0
5. Set Bus count p = 1
6. if p = s
   Go to step 11
   else
   Calculate the real and reactive powers Pp and Qp respectively
7. Evaluate ΔP p k = P psp – Pp k
8.   if p = generator bus
   Compare Qp k with the limits
   if Qp<Qp min
   Set Qp = Qpmin
   else
      if Qp>Qp max
   Set Qp = Qpmax; treat the bus as load bus
   else
   go to step 9
      endif
   endif
   else
Evaluate voltage residues & go to step 11
endif

8.   Evaluate $\Delta Q\,p\,k = Q\,ps - Qp\,k$

9.   Set $p = p + 1$
endif
10. if $p > n$
        Determine the largest of the absolute value of residue
   else
        Go to step 6
   endif
11. if residue $\leq \varepsilon$
        Go to step 15
   else
        Evaluate elements for Jacobian matrix
   endif
12. Calculate voltage increments Δepk and $\Delta f\,pk$
13. Evaluate $\cos\delta$ and $\sin\delta$ of all voltages.
14. if p = generator bus?
        Calculate $e\,p\,k+1$ and $f\,p\,k+1$ for generator bus
   else
        if $|Vp| < |Vp\,min|$
                Calculate $e\,p\,k+1$ and $f\,p\,k+1$ for minimum condition.
        else

        if $|Vp| > |Vp\,max|$
                Calculate $e\,p\,k+1$ and $f\,p\,k+1$ for maximum condition.
        else
        Advance iteration count $K = K + 1$ and go to step 5
   15. Evaluate bus and line powers and print the results.

5.   Results and Discussions

IEEE-6 bus consists of 3 synchronous generators with IEEE type-1 exciters, 2 of which are synchronous compensators used only for reactive power support. There are total load of 40 MW and 18 Mvar distributed in 4 load buses. Bus 1 is slack bus which is denoted by Type 1; Bus 2 is Generating Bus which is denoted by Type 2; Bus 3, Bus 4, Bus 5 & Bus 6 are the load busses which are denoted by Type 0.

Load flow results for IEEE6 bus system shows that the weakest bus on the system is Bus – 3. It is required now to determine the point of collapse where the application of SVC can be implemented to improve and check for voltage stability. Load flow analysis for the system is performed with changing load. It was found at Bus – 3 that the voltage magnitude is drooping in characteristic as can be seen from the Table 1. Figure 1 show the drooping plot of PV curve for load increased to 120 MW.

On further increasing the load, the drop in the magnitude of voltage is being steeper. At the load of 129 MW, the load flow solution did not converge; still giving a solution in the range of voltage drop & not in the range of voltage collapse. Also it means that on Bus 3 maximum loadability is around 129 MW. When load is increased to 129 MW, voltage magnitude of Bus 3 was 0.625pu and when the load is 129.500 MW, voltage magnitude of Bus 3 reaches to -0.743 pu. The plot of the PV curve is shown in Figure 2.

After the load of 129.5 MW, Bus 3 become voltage magnitude becomes negative so it will be the point of collapse at this condition. Compensation can be applied on the maximum load (or collapse point) to achieve voltage stability.

Table 1. The voltage magnitude at Bus – 3

| | LOAD | |
Pg (MW)	Qg (Mvar)	Voltage Mag.
10	6.5	0.956
15	6.5	0.953
20	6.5	0.949
25	6.5	0.946
30	6.5	0.941
35	6.5	0.937
40	6.5	0.932
45	6.5	0.927
50	6.5	0.921
55	6.5	0.915
60	6.5	0.909
65	6.5	0.902
70	6.5	0.894
75	6.5	0.885
80	6.5	0.876
85	6.5	0.866
90	6.5	0.855
95	6.5	0.843
100	6.5	0.829
105	6.5	0.814
110	6.5	0.796
115	6.5	0.775
120	6.5	0.748

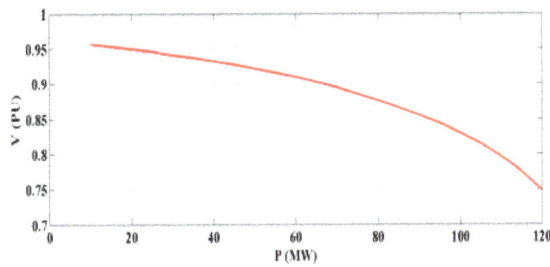

Figure 1. The drooping plot of PV curve for load increased to 120 MW

Table 2. The voltage magnitude of maximum loadability

| | LOAD | |
Pg (MW)	Qg (Mvar)	Voltage Mag.
125	6.5	0.709
126	6.5	0.698
127	6.5	0.685
128	6.5	0.666
128.5	6.5	0.649
129	6.5	0.625
129.5	6.5	-0.743

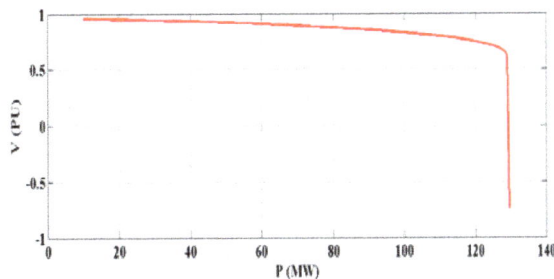

Figure 2. The plot of the PV curve when the load is 129.500 MW

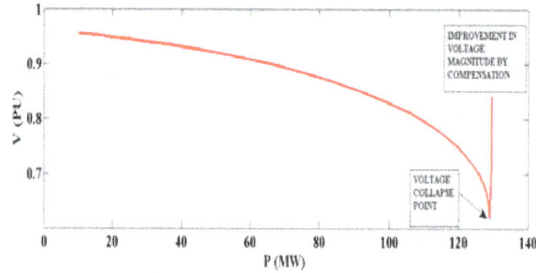

Figure 3. The plot of the PV curve with SVC compensating 25 Mvar

As seen from the plot in Figure 3 at the very same point of voltage collapse; at a load of 129.5 MW; compensating with SVC with 25 Mvar; voltage magnitude has again reached in the stable region. Also it is known that applying compensation above 30 % becomes costlier than implementing a new system; hence this is the maximum stability that can be achieved through implementing SVC on IEEE 6 Bus System on maximum load to avoid voltage collapse.

Table 3

LOAD		Voltage Mag.	Injected Mvar
Pg (MW)	Qg (Mvar)		
129.5	6.5	0.714	5
129.5	6.5	0.757	10
129.5	6.5	0.79	15
9.5	6.5	0.818	20
129.5	6.5	0.843	25

From the study it can be concluded the point of maximum loading to be 129.5 MW of load on the weakest bus; Bus 3 and also the optimum voltage stability through implementation of SVC is obtained; that is with application of SVC with reactive injection of Mvar of 25 MW the magnitude of voltage reaches to 0.843 PU from -0.743 PU.

6. Conclusion & Future Scope

From the study it can be concluded the point of maximum loading to be 129.5 MW of load on the weakest bus; Bus 3 and also the optimum voltage stability through implementation of SVC is obtained; that is with application of SVC with reactive injection of Mvar of 25 MW. The magnitude of voltage reaches to 0.843 PU from -0.743 PU.

The study has evaluated the maximum load before voltage collapse on a IEEE 6 bus system, the same work can be performed with different standard as well as practical systems to verify and implement it to the existing networks.SVC on its standard rating is been applied here to observe voltage stability, Parameter of SVC can be altered to improve for more voltage stability condition. This study is applied only to AC test System; it can be very well implied on HVDC system and can be verified for the same. SVC is been applied to the test system, there are various other FACTS devices available for stability analysis, like STATCOM, UPFC, SSSC, TCSC etc which individually have different characteristics. Other than FACTS devices, there are various other electronic devices which can be implemented in the same system like OLTC, Tap-changing transformers. Artificial Intelligence & Fuzzy Logic has emerged as techniques which can be implemented to many realistic power system networks in its various domains. There are various methodologies available which can used to extend this work, through which the weakest bus for any larger system can be evaluated very fastly and accurately.

References

[1] Brit KA, Graff JJ, McDonald JD, El-Abiad AH, Three phase load flow program. *IEEE Transactions on Power Application System*. 1976; 95(1): 59-65.
[2] Mamdouh Abdul Akher, Khalid Mohamed Nora, Abdul Halim Abdul Rachid. Improved three phase power flow methods using sequence components. *IEEE Transactions on Power systems*. 2005; 20(3): 1389-1397.
[3] Yamile E-Del Valle. Optimization of power system performance using FACTS Devices. *IEEE transactions on Power Systems*. 2009; 62(2).
[4] Jing Zhang, JY Wen, SJ Cheng, Jia Ma. A Novel SVC allocation method for power system voltage stability enhancement by normal forms of diffeomorphism. *IEEE transactions on Power Systems*. 2007; 22(4): 1818-1825.
[5] Massoud Amin. *Modernizing The National Electric Power Grid*. Workshop on Modernizing the National Electric Power Grid EPRI/NSF/ Entergy and DOE. New Orleans, LA. 2002; 18-19.
[6] Chen-Ching Liu, James Momoh and Paul Werbos, Massoud Amin, Aty Edris, and Acher Mosse. NSF/EPRI Workshop on Urgent Opportunities for Transactions on mission System Enhancement, EPRI. Palo Alto, CA. 2012; 11-12.
[7] Clark W Gellings. Power Delivery System of the Future. *IEEE Power Engineering Review*. 2002; 22(12): 7-12.
[8] Gregory J Miranda. *Options for the Cement Industry in the Deregulated Power Era*. Cement Industry Technical Conference, In Proceedings of 44th IEEE-IAS/PCA. 2002; 19-28.
[9] C Mensah-bonsu, S Oren. California Electricity Market Crisis: Causes, Remedies, and Prevention. *IEEE Power Engineering Review*. 2002; 22(8) 4-23.
[10] Brian Cory, Peter Lewis. *The Reorganization of the Electric Supply Industry – A critical Review*. Power Engineering Journal. 1997; 42: 42-46.
[11] Massoud Amin. *Evolving Energy Enterprise—"Grand Challenges": Possible road ahead and challenges for R&D*. IEEE Power Engineering Society Summer Meeting. 2002; 1705-1707.
[12] Narain G Hingorani, Laszlo Gyugyi. Understanding FACTS. *IEEE Press*. New York: 2000: ISBN 0-7803-3455-8.
[13] Kundur P. Power System Stability and Control. 1st Edtion. NewYork: Tata McGraw-Hill Education. 1994: ISBN-10: 0070635153.
[14] Ajjarapu V. Computational Techniques for Voltage Stability Assessment and Control. 1st Edition. Springer, New York: 2006: ISBN-10: 0387260803.
[15] Mathur RM, RK Varma. Thyristor-Based Facts Controllers for Electrical Transmission Systems. 1st Edition. New York: IEEE Press. 2002: ISBN- 10: 0-4712-0643-1.
[16] Sode-Yome, AN Mithulananthan, KY Lee. *Static voltage stability margin enhancement using STATCOM, TCSC and SSSC*. Proceedings of IEEE/PES Transmission and Distribution Conference and Exhibition: Asia and Pacific, IEEE Xplore Press Dalian. 2005; 1-6.
[17] Natesan R, G Radman. *Effects of STATCOM, SSSC and UPFC on voltage stability*. Proc. 36th Southeastern Symposium on System Theory, IEEE Xplore Press. 2004; 546-550.
[18] Gotham DJ, GT Heydt. Power Flow control and Power Flow Studies for Systems with FACTS devices. *IEEE Transactions on. Power System*. 1998; 13(4): 60-65.
[19] Acha E. Facts: Modelling and Simulation in Power Networks. 1st Edition. Chichester: John Wiley and Sons. 2004: ISBN-10: 0470852712.
[20] Hammad, B Roesle. New Roles for Static VAR Compensators in Transmission Systems. Brown Boveri Review. 1986; 73: 314-320.

25

Facial Expression Recognition using SVM Classifier

Vasanth P.C*, Nataraj K.R
ECE Department, SJB Institute of Technology, Bangalore, India
e-mail: Pcvasanth10@gmail.com

Abstract

Facial feature tracking and facial actions recognition from image sequence attracted great attention in computer vision field. Computational facial expression analysis is a challenging research topic in computer vision. It is required by many applications such as human-computer interaction, computer graphic animation and automatic facial expression recognition. In recent years, plenty of computer vision techniques have been developed to track or recognize the facial activities in three levels. First, in the bottom level, facial feature tracking, which usually detects and tracks prominent landmarks surrounding facial components (i.e., mouth, eyebrow, etc), captures the detailed face shape information; Second, facial actions recognition, i.e., recognize facial action units (AUs) defined in FACS, try to recognize some meaningful facial activities (i.e., lid tightener, eyebrow raiser, etc); In the top level, facial expression analysis attempts to recognize some meaningful facial activities (i.e., lid tightener, eyebrow raiser, etc); In the top level, facial expression analysis attempts to recognize facial expressions that represent the human emotion states. In this proposed algorithm initially detecting eye and mouth, features of eye and mouth are extracted using Gabor filter, (Local Binary Pattern) LBP and PCA is used to reduce the dimensions of the features. Finally SVM is used to classification of expression and facial action units.

Keywords: *face expression and recognition, LBP, PCA, Support vehicle machine, Gabor feature*

1. Introduction

Facial feature points encode critical information about face shape and face shape deformation. Accurate location and tracking of facial feature points is important in the applications such as animation, computer graphics, etc. Generally, the facial feature points tracking technologies could be classified into two categories: model free and model-based tracking algorithms. Model free approaches are general purpose point trackers without the prior knowledge of the object. The second type is Model based methods, such as active shape model (ASM), active appearance model (AAM), direct appearance model (DAM), etc, on the other hand, focus on explicit modelling the shape of the objects.

Facial expression recognition systems usually try to recognize either six expressions or the (Action Units) AUs. Over the past decades, there has been extensive research in computer vision on facial expression analysis. Current methods in this area can be grouped into two categories: image-driven method and model-based method. Image-driven approaches, which focus on recognizing facial actions by observing the representative facial appearance changes, usually try to classify expression or AUs independently and statically. Model-based methods make use of the relationships among AUs, and recognize the AUs simultaneously.

The proposed project work consists of two phases training and testing phase. In training phase extracting features of the eye and mouth and creating the knowledge base. At the time of testing make use of knowledge base extracting Face expressions and action units. The proposed project consists of fallowing blocks face localization, eye and mouth detection, Features extraction and classification. Face localization is a crucial first step for proposed project, here image is input to this block, and first we exploit the colour information to limit the search area to candidate eye and mouth regions. In second block determine the exact eye and mouth position. After detecting eye and mouth region, extracting two types of features of the interested region using Gabor filter and LBP. Features are extracted directly from gray-scale character images by Gabor filters which are specially designed from statistical information of character structures. LBP is a simple yet very efficient texture operator which labels the pixels of an image by Thresholding the neighbourhood of each pixel and considers the result as a binary number.

Due to its discriminative power and computational simplicity, LBP texture operator has become a popular approach in various applications. It can be seen as a unifying approach to the traditionally divergent statistical and structural models of texture analysis. Perhaps the most important property of the LBP operator in real-world applications is its robustness to monotonic gray-scale changes caused, for example, by illumination variations. Another important property is its computational simplicity, which makes it possible to analyze images in challenging real-time settings. The extracted features using Gabor and LBP the features size is large, PCA is used to reduce the features size of eye and mouth. These extracted features are stored in knowledge base for testing purpose. In testing phase applying above method to extract the features and classifying the extracted features using SVM.

2. Methodology

The proposed algorithm methods includes the following modules:
a)   Gabor Features
b)   LBP Features
c)   Principal component analysis
d)   Support Vehicle Machine

The project work consists of two phases training and testing phase. In training phase extracting features of the eye and mouth and creating the knowledge base. At the time of testing make use of knowledge base extracting Face expressions and action units. The proposed project consists of fallowing blocks face localization, eye and mouth detection, Features extraction and classification. Face localization is a crucial first step for proposed project, here image is input to this block, and first we exploit the colour information to limit the search area to candidate eye and mouth regions. In second block determine the exact eye and mouth position. After detecting eye and mouth region, extracting two types of features of the interested region using Gabor filter and LBP. The extracted features using Gabor and LBP the features size is large, PCA is used to reduce the features size of eye and mouth. These extracted features are stored in knowledge base for testing purpose. In testing phase applying above method to extract the features and classifying the extracted features using SVM
.

2.1. Gabor Features

Gabor features constructed from post-processed Gabor filter responses have been successfully used in various important computer vision tasks, such as in texture segmentation, face detection, and iris pattern description. However, only very rarely the main weakness of Gabor filter based features, the computational heaviness, has received any attention even though it may prevent the use of proposed methods in real applications. It is evident that Gabor filters have many advantageous or even superior properties for feature extraction, but if the computational complexity cannot be improved their application areas will remain limited.

Some properties of Gabor filters:

A tunable bandpass filter. Similar to a STFT or windowed Fourier transform. Satisfies the lower-most bound of the time-spectrum resolution (uncertainty principle).

It a multi-scale, multi-resolution filter. Has selectivity for orientation, spectral bandwidth and spatial extent. Has response similar to that of the Human visual cortex (first few layers of brain cells). Used in many applications – texture segmentation; iris, face and fingerprint recognition. Computational cost often high, due to the necessity of using a large bank of filters in most applications.

2.2. LBP Features

LBP is a simple yet very efficient texture operator which labels the pixels of an image by Thresholding the neighbourhood of each pixel and considers the result as a binary number. Due to its discriminative power and computational simplicity, LBP texture operator has become a popular approach in various applications. It can be seen as a unifying approach to the traditionally divergent statistical and structural models of texture analysis. Perhaps the most important property of the LBP operator in real-world applications is its robustness to monotonic gray-scale changes caused, for example, by illumination variations. Another important property is its computational simplicity, which makes it possible to analyze images in challenging real time settings.

2.3. Principal Compnent Analysis

Principal component analysis (PCA) is possibly the dimension reduction technique most widely used in practice, perhaps due to its conceptual simplicity and to the fact that relatively efficient algorithms (of polynomial complexity) exist for its computation. In signal processing it is known as the Karhunen -Loeve transform.

a) PCA is a powerful and widely used linear technique in statistics, signal processing, image processing, and elsewhere.

b) Several names: the (discrete) Karhunen- Loève transform (KLT, after Kari Karhunen and Michael Loève) or the Hotelling transform (after Harold Hotelling).

c) In statistics, PCA is a method for simplifying a multidimensional dataset to lower dimensions for analysis, visualization or data compression.

d) PCA represents the data in a new coordinate system in which basis vectors follow modes of greatest variance in the data.

e) Thus, new basis vectors are calculated for the particular data set

f) The price to be paid for PCA's flexibility is in higher computational requirements as compared to, e.g., the fast Fourier transform.

2.4. Support Vehicle Machine

Support Vector Machine (SVM) was first heard in 1992, introduced by Boser, Guyon, and Vapnik in COLT-92. Support vector machines (SVMs) are a set of related supervised learning methods used for classification and regression. They belong to a family of generalized linear classifiers. In another terms, Support Vector Machine (SVM) is a classification and regression prediction tool that uses machine learning theory to maximize predictive accuracy while automatically avoiding over-fit to the data. Support Vector machines can be defined as systems which use hypothesis space of a linear functions in a high dimensional feature space, trained with a learning algorithm from optimization theory that implements a learning bias derived from statistical learning theory. Support vector machine was initially popular with the NIPS community and now is an active part of the machine learning research around the world. SVM becomes famous when, using pixel maps as input; it gives accuracy comparable to sophisticated neural networks with elaborated features in a handwriting recognition task [2]. It is also being used for many applications, such as hand writing analysis, face analysis and so forth, especially for pattern classification and regression based applications. The foundations of Support Vector Machines (SVM) have been developed by Vapnik.

3. Results and Discussion

The results are obtained using a MATLAB simulator tool which is licensable software. The simulation results are shown below.

The results which shows face recognition it first creates database of expressions, secondly SVM training of input image then select an query of image then it localizes eyes and mouth and finally face expression recognition is done.

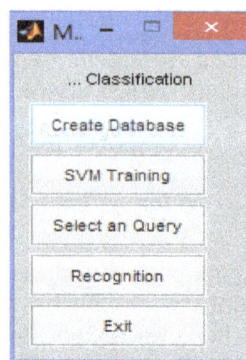

The above figure shows Query image in which it first face detection is done.

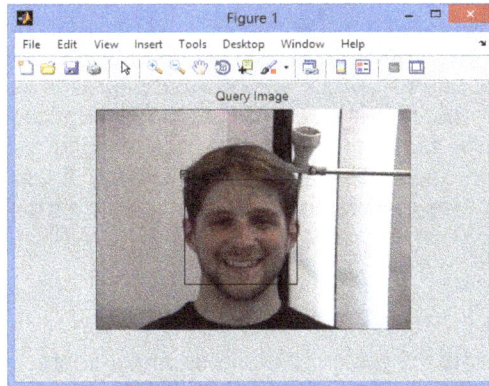

Figure 1. Query image

The above fig shows The figure shows the recognition of eyes and mouth once face detected then exploit color information to limit the search area to candidate eyes and mouth regions. The Gabor and LBP features are used for cropping of eyes and mouth from the face region.

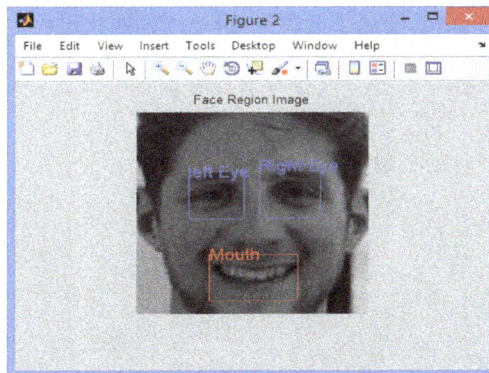

Figure 2. Detecting eyes and mouth

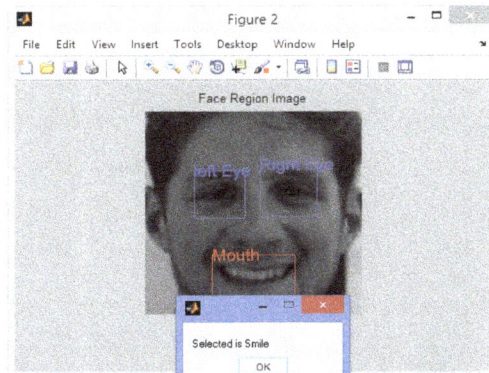

The figure shows face expression recognition. In the dataset four expressions are given in training phase .The eyes and mouth are detected through the features as the features is large PCA is used to reduce the feature of eyes and mouth the SVM is used for classification of the expression once it is classified then recognition of the expression is done. The above figure shows the recognized expression which is selected as smile.

4. Conclusion

In this project, we have proposed a combination of two methods Gabor and LBP feature to extract the features of eye and mouth. In which Gabor filter which are specially designed from statistical information. LBP operator is used in real world applications then applying PCA to reduce the dimensions of the features matrix that is eyes and mouth. Small size of feature matrix helps to increase the speed of the classification and finally using SVM the extracted features are classified, the classified features have good accuracy displaying the expression and facial action units.

Acknowledgements

I would like to articulate my profound gratitude and indebtedness to my project guide Dr. Nataraj. K.R (Professor & HOD Dept. of ECE, SJBIT), who is always being a constant motivation and guiding factor throughout the project time in and out as well. It has been a great pleasure for me to an opportunity to work under him.

References

[1] Z Zhu, Q Ji, K Fujimura, K Lee. *Combining Kalman filtering and mean shift for real time eye tracking under active IR illumination.* In Proc. IEEE Int. Conf. Pattern Recognit. 2002; 4: 318–321.

[2] SJ McKenna, S Gong, RP Würtz, J Tanner, D. Banin. *Tracking facial feature points with Gabor wavelets and shape models.* In Proc. Int. Conf. Audio-Video-Based Biometric Person Authent. 1997; 1206: 35–42.

[3] Y Tong, Y Wang, Z Zhu, Q Ji. Robust facial feature tracking under varying face pose and facial expression. *Pattern Recognit.* 2007; 40(11): 3195–3208.

[4] J Whitehill, CW Omlin. *Haar features for FACS AU recognition.* In Proc. IEEE Int. Conf. Autom. Face Gesture Recognit. 2006; 97–101.

[5] L Zhang, D Tjondronegoro. Facial expression recognition using facial movement features. *IEEE Trans. Affect. Comput.* 2011; 2(4): 219–229.

Permissions

The contributors of this book come from diverse backgrounds, making this book a truly international effort. This book will bring forth new frontiers with its revolutionizing research information and detailed analysis of the nascent developments around the world.

We would like to thank all the contributing authors for lending their expertise to make the book truly unique. They have played a crucial role in the development of this book. Without their invaluable contributions this book wouldn't have been possible. They have made vital efforts to compile up to date information on the varied aspects of this subject to make this book a valuable addition to the collection of many professionals and students.

This book was conceptualized with the vision of imparting up-to-date information and advanced data in this field. To ensure the same, a matchless editorial board was set up. Every individual on the board went through rigorous rounds of assessment to prove their worth. After which they invested a large part of their time researching and compiling the most relevant data for our readers.

The editorial board has been involved in producing this book since its inception. They have spent rigorous hours researching and exploring the diverse topics which have resulted in the successful publishing of this book. They have passed on their knowledge of decades through this book. To expedite this challenging task, the publisher supported the team at every step. A small team of assistant editors was also appointed to further simplify the editing procedure and attain best results for the readers.

Apart from the editorial board, the designing team has also invested a significant amount of their time in understanding the subject and creating the most relevant covers. They scrutinized every image to scout for the most suitable representation of the subject and create an appropriate cover for the book.

The publishing team has been an ardent support to the editorial, designing and production team. Their endless efforts to recruit the best for this project, has resulted in the accomplishment of this book. They are a veteran in the field of academics and their pool of knowledge is as vast as their experience in printing. Their expertise and guidance has proved useful at every step. Their uncompromising quality standards have made this book an exceptional effort. Their encouragement from time to time has been an inspiration for everyone.

The publisher and the editorial board hope that this book will prove to be a valuable piece of knowledge for researchers, students, practitioners and scholars across the globe.

List of Contributors

Kunal Chakraborty, Indranil Roy and Palash De
Department of Electrical Engineering, IMPS College of Engineering & Technology Malda, West Bengal, 732103, India

Azhar AbduAlwahab Ali, K. T. Al-Rasoul and Issam M. Ibrahim
Iraqi Ministry of Sciences and Technology, Iraq

Bushra R.Mhdi, Gaillan H.Abdullah, Nahla A.Aljabar and Basher R.Mhdi
Ministry of Science and Technology, Iraq, Bagdad

K. Lenin, B. Ravindhranath Reddy and M. Surya Kalavathi
Jawaharlal Nehru Technological University Kukatpally, Hyderabad 500 085, India

K. Lenin, B. Ravindhranath Reddy and M. Suryakalavathi
Jawaharlal Nehru Technological University Kukatpally, Hyderabad 500 085, India

D. V. N Ananth
Viswanadha Institute of Technology and Management, Visakhapatnam

G. V. Nagesh Kumar
GITAM University, Visakhapatnam

Gunawan Dewantoro, Deddy Susilo and Ditya Clarisa Amanda
Department of Electronics and Computer Engineering, Satya Wacana Christian University, 52-60 Diponegoro Street, Salatiga, Indonesia 50711

Satyam Gupta and Gunjan Gupta
Department of Electronics and Communication Engineering, Invertis University, Bareilly, India

Saurabh Mishra
Department of Electronics and Communication Engineering Invertis University, Bareilly, Uttar Pradesh 243123, India

Omid Alavi and Behzad Vatandoust
Department of Electrical Engineering, K.N. Toosi University of Technology, Tehran, Iran

Muhammad Ali Raza Anjum
Department of Electrical Engineering, Army Public College of Management and Sciences, Rawalpindi, Pakistan

Md. Imran Azim, Md. Abdul Wahed and Md. Ahsanul Haque Chowdhury
Departement of Electrical and Electronic Engineering, Rajshahi University of Engineering and Technology (RUET)

J. Manikandan, C.S. Celin and V.M. Gayathri
Department Of Computer Science and Engineering, Saveetha School of Engineering, Saveetha University, Chennai

Jafar Mostafapour
Azerbaijan Regional Electric company, Tariz, Iran

Murtaza Farsadi
Department of Electrical Engineering, Urmia University

Jafar Reshadat
Azerbaijan Regional Electric company, Tariz, Iran Department Management, Science and Research Branch Islamic Azad University, West Azerbaijan, Iran

A.S Kang
ECE Deptt, Panjab University Regional Centre, Hoshiarpur, Punjab, INDIA

Renu Vig
ECE Deptt, Panjab University, Chandigarh, INDIA

Hossein Sadegh Lafmejani and Hassan Zarabadipour
Faculty of Technical and Engineering, Imam Khomeini International University, Qazvin, Iran

Bhupendra Sehgal, S P Bihari and Yogita Kumari
Inderprastha Engineering College, Ghaziabad, India

R.N.Chaubey and Anmol Gupta
KIET Ghaziabad, India

K. Lenin
Research Scholar, JNTU, Hyderabad 500 085 India

B. Ravindhranath Reddy
Deputy Executive Engineer, JNTU, Hyderabad 500 085 India

M. Surya Kalavathi
Department of Electrical and Electronics Engineering, JNTU, Hyderabad 500 085, India

G.Vasumathi and P.Subashini
Department of Computer Science, Avinashilingam University, Coimbatore, India

V.Balaji
Department of Electrical Engineering, Singhania University, Pacheri Bari, Rajasthan, India

Dr. L.Rajaji and Shanthini K.
ARM College of Engineering and Technology, MaraiMalai Nagar, Chennai, India

Shamik Tiwari
CET, Mody University of Science & Technology, Lakshmangarh

Tesfaye Alamirew and Nigus Gabbeye
Faculty of Chemical and Food Engineering, BIT, Bahir Dar University, Ethiopia

V. Balaji
Faculty of Electrical and Computer Engineering, BIT, Bahir Dar University, Ethiopia

A. Bennouk, A. Nejmi and M. Ramzi
Mathematics and physics laboratory Faculty of Sciences and Technics, Beni Mellal Sultan Moulay Slimane University, Beni Mellal, Morocco

Sita Singh, Jitendra Hanumant and Ashutosh Kashiv
Electrical & Electronics Engineering Department, Oriental University Opposite Revati Range Gate no. 1, Sanwer Road, Jakhya, Indore, MP, India 453555

Vasanth P.C and Nataraj K.R
ECE Department, SJB Institute of Technology, Bangalore, India

Index

www.ingramcontent.com/pod-product-compliance
Lightning Source LLC
Chambersburg PA
CBHW082028190326
41458CB00010B/3300